ENCYCLOPAEDIA OF EMINENT THINKERS

VOLUME-XVIII: **K.B. HEDGEWAR**

Volumes by the same author

The Political Thought of ***Veer Savarkar***
The Political Thought of ***M.A. Jinnah***
The Political Thought of ***Mohammad Iqbal***
The Political Thought of ***K.B. Hedgewar***
The Political Thought of ***M.S. Golwalkar***
The Political Thought of ***Lokmanya Bal Gangadhar Tilak***
The Political Thought of ***Pandit Deendayal Upadhyaya***
The Political Thought of ***Gopal Krishna Gokhle***
The Political Thought of ***Maulana Abul Kalam Azad***
The Political Thought of ***Sir Syed Ahmed Khan***

About the Author

Dr. Jai Narain Sharma (*b.* 1951) is Professor and Chairman of the Department of Gandhian Studies, Panjab University, Chandigarh. He is also Hon. Director of Gandhi Bhawan; member of Board of Studies, Nagpur University, Nagpur and Mahatma Gandhi Kashi Vidyapith, Varanasi; member of Research Degree Committee of the Department of Political Science, H.P. University, Shimla: General Secretary of Indian Soviety of Gandhian Studies; and member on the Presidium of the Alliance for Sarvodaya. He regularly writes for leading newspapers, and has published more than hundred research papers in professional journals of repute. Dr. Sharma has many books to his credit including *"Gandhi's View of Political Power", "Economics of Defence: A Study of SAARC Countries", "Economic Thought of Mahatma Gandhi", "Human Resource Management", "Alternative Economics: Economics of Mahatma Gandhi and Globalisation", "Power, Politics and Corruption: A Gandhian Solution"* and *"Research Methodology: The Discipline and its Dimensions.*

ENCYCLOPAEDIA OF EMINENT THINKERS

VOLUME EIGHTEEN

POLITICAL THOUGHT OF K.B. HEDGEWAR

Jai Narain Sharma
Professor and Chairman
Department of Gandhian Studies
Panjab University, Chandigarh

CONCEPT PUBLISHING COMPANY, NEW DELHI-110059

Cataloging in Publication Data—DK
Courtesy: D.K. Agencies (P) Ltd. <docinfo@dkagencies.com>

Encyclopaedia of eminent thinkers / K.S. Bharathi.
v. 16-20 cm. — (Gandhian studies and peace research series ; 11)
Vols. 16, 19 by Ashu Pasricha; v. 17-18 & 20 by Jai Narain Sharma.
Includes bibliographical references.
Includes indexes.
ISBN 13: 9788170226840 (set)
ISBN 10: 8170226848 (set)

1. Statesmen—India. 2. Political science—India. 3. India—Politics and government. I. Bharathi, K. S. (Kuduva Swamy), 1956- II. Pasricha, Ashu, 1967- III. Sharma, Jai Narain, 1951- IV. Series: Gandhian studies and peace research series ; 11.

DDC 954.035 22

ISBN-13: 978-81-7022-684-0 (Series)
ISBN-10: 81-7022-684-8 (Series)
ISBN-13: 978-81-8069-498-1 (Vol. XVIII)
ISBN-10: 81-8069-498-4 (Vol. XVIII)

First Published 2008

Published and Printed by
Ashok Kumar Mittal
Concept Publishing Company
A/15-16, Commercial Block, Mohan Garden
New Delhi-110059 (India)
Phones: 25351460, 25351794
Fax: 091-11-25357103
E-mail: publishing@conceptpub.com

DEDICATED
TO
THE SACRED MEMORY
OF MY
RESPECTED FATHER
(Late) PANDIT SADHU RAM VAIKARANACHRIYA
(1913-2005)

PREFACE

Independent India is engaged in a many-sided renaissance and reconstruction more or less consciously directed to what may compendiously be called nation making (or re-making) in an image more adequate to the needs of the present and future of India as a strong, creative nation guiding the destinies of its people in accordance with a worthy and inspiring charter of life.

This era commenced roughly with the emergence and work of Raja Rammohun Roy in the beginning of the nineteenth century and has continued to the present day through the life-effort of notable thinkers in many fields, like Swami Vivekananda, Swami Dayananda, Shri Aurobindo, Lokmanya Tilak, Mahatma Gandhi, Gurudev Tagore, Veer Savarkar and Dr. Hedgewar. These are only typical names and do not exhaust the galaxy.

In the phase of the struggle (and triumph) with the British Power (typifying the entire gamut of foreign ideas and ideals), we have a similar period as in many periods in the past of the self-recollection and re-assertion of the national self-consciousness and of a conscious search for the roots of our culture as the living points on which we may re-graft current life and foster it to vigorous growth and power. In this many-sided effort, the image of Indian culture as the pattern of nation building is laid over with many confused notions from the West and from distorted ideas of the past of our own life.

It is too often forgotten that Hinduism is not merely a sect, a small religious fellowship concerned exclusively with modes of worship or social customs peculiar to it. The word Hindu in this context has a national character. It is tantamount to the word Indian, i.e. pertaining to a people living beside the river Sindhu. It connotes the entire culture and civilisation of the Indian people from prehistoric times developed on Indian soil through millennia.

It is in this context Dr. Hedgewar preached a more positive and more creative policy with a clear vision of the best features of Indian society at its best and of the enduring element of nationalism and Westernism in just proportion and with sure insight to their human significance.

Hedgewar was a born patriot, and a fiery one at that. Even as a boy of tender age, sparks of protest against the foreign British rule flew from him. He braved rustication from the school, shadowing by the government informers and hardships to his uncle. He set out to Calcutta more to delve into the intricacies of the revolutionary movement than to study medicine. Back home as a doctor, he turned his back on personal and family happiness and plunged into the mass movements for *Swaraj* led by Congress at the time. Twice he courted imprisonment completing hard terms in prison.

Hedgewar's speeches were marked for their fiery character. Listening to the Doctor's defiant statement in the court in 'defence' of one of his public speeches for which he was hauled up for 'sedition', the judge remarked, "Your defence is more seditious than your speech" and awarded him one year's rigorous imprisonment. That was in 1921. The 1930 movement again saw him courting imprisonment by offering *Satyagraha* this time with some of his colleagues in the Sangh also joining him.

The desire to see the country free in his own lifetime and with his own eyes remained a flaming passion with him till his very end in 1940. Those were the days of the Second World War when the British were caught in a life-and-death struggle. Hedgewar had sensed beforehand the gathering war clouds and felt that the enslaver's difficulty could prove an opportunity to the enslaved to shake off the shackles of slavery. He urged all the country men workers to speed up the work and he himself toiled day and night to build a mighty force in the shortest possible time.

The present work is an humble attempt to outline the political ideas of this splendid personality. I hope the book will provide a glimpse—though not a full effulgent view of the Himalayan height of the personality of Dr. Hedgewar.

Personal acknowledgments are no easy matters where so many have been kind. My special thanks are due to Professor S.L. Malhotra, Professor R.N. Pal, Professor N. Radhakrishnan, Dr. K.S. Bharathi, Dr. R.P. Dwivedi, Dr. A.D. Mishra, Professor B.R. Duggar, Professor Raj Bir Singh Yadav, Professor M.L. Sharma, Dr. Meena Dutta, Dr. Ashu Pasricha and Dr. Rashmi Sharma. Beyond that I can make only an all embracing salute to the galaxy of the scholars whose works I have consulted.

I am also grateful to Shri Ashok Kumar Mittal, my publisher, for his consistent zeal in the printing and publication of this volume.

Chandigarh **JAI NARAIN SHARMA**

Encyclopaedia of Eminent Thinkers Series

Vol. I. The Political Thought of **Mahatma Gandhi**/ *K.S. Bharti*
Vol. II. The Political Thought of **Nehru**/ *K.S. Bharti*
Vol. III. The Political Thought of **Rabindra Nath Tagore**/ *K.S. Bharti*
Vol. IV. The Political Thought of **Radhakrishnan**/ *K.S. Bharti*
Vol. V. The Political Thought of **Vinoba**/ *K.S. Bharti*
Vol. VI. The Political Thought of **Aurobindo**/ *K.S. Bharti*
Vol. VII. The Political Thought of **Jayaprakash Narayan**/ *K.S. Bharti*
Vol. VIII. The Political Thought of **Vivekananda**/ *K.S. Bharti*
Vol. IX. The Political Thought of **Ambedkar**/ *K.S. Bharti*
Vol. X. The Political Thought of **M.N. Roy**/ *K.S. Bharti*
Vol. XI. The Political Thought of **Dadabhai Naoroji**/ *Ashu Pasricha*
Vol. XII. The Political Thought of **Veer Savarkar**/ *Jai Narain Sharma*
Vol. XIII. The Political Thought of **M.A. Jinnah**/ *Jai Narain Sharma*
Vol. XIV. The Political Thought of **Tej Bahadur Sapru**/ *Ashu Pasricha*
Vol. XV. The Political Thought of **C. Rajagopalachari**/ *Ashu Pasricha*
Vol. XVI. The Political Thought of **Subhas Chandra Bose**/ *Ashu Pasricha*
Vol. XVII. The Political Thought of **Mohammad Iqbal**/ *Jai Narain Sharma*
Vol. XVIII. The Political Thought of **K.B. Hedgewar**/ *Jai Narain Sharma*
Vol. XIX. The Political Thought of **Bhagat Singh**/ *Ashu Pasricha*
Vol. XX. The Political Thought of **M.S. Golwalkar**/ *Jai Narain Sharma*

CONTENTS

1

LIFE AND ACTIVITIES

Hedgewar family originally hailed from Kandkurti village in Telangana, a village with a population of just over two thousand. Kandkurti is in the Bodhana tehsil in the Indore (Nizamabad) district, situated on the border between Andhra and Maharashtra. Near the village is the sacred confluence of Godavari, Vanjar and Haridra rivers. The hallowed Vanjar confluence finds mention even in the *puranas*. As if reflecting this *sangam*, one finds in this area a delightful admixture of three languages Kannada, Telugu and Marathi.

The places were at one time the abode of scholars and prosperous Brahmin families. The Hedgewar family was one such. They were Deshastha Brahmins of the Shakala branch, belonging to the Ashwalayana Sutra of the Rigveda. Their gotra was Kashyapa, and learning and transmission of the *Vedas* was their sole preoccupation. The *agnihotra* too was in vogue. There are documents indicating that in the course of his travels, Shankaracharya used to nominate someone from the Hedgewar family to be in-charge of propagation of *dharma* in this area.[1] Many are the families claiming to be followers of the Hedgewar, and their family documents speak with utmost reverence of the Hedgewar saying, "*Hedage Kulaguru poorvapara, jaise suryavamsha vasisthavara*" ("The tradition of Hedge *kulagurus* is (glorious) like that of Vasishtha in the solar dynasty.")[2]

In the beginning of nineteenth century, many Brahmin families left the Telangana region owing to the neglect and penury suffered by them under the Mughals. Several such families chose to settle in

Nagpur since the Bhonsle rulers were great patrons of Vedic learning. Among them was Narahara Shastri, the great grandfather of Baliram Pant of the Hedgewar clan. In 1853, Nagpur came under the yoke of the British rulers, and English education gained prominence to the utter neglect of traditional learning. Great scholars, well versed in many *shastras* had to resort to priestcraft to earn their livelihood. Even under such hard conditions, Baliram Pant Hedgewar sustained the family tradition with efficiency and erudition. His wife Revatibai came from the Paithankar family, and was known for her serenity and amiable temperament. Despite poverty, the couple spent their days in peace and happiness.

It was in such peaceful environment that Keshav was born, on 1st April, 1889. Keshav was the fifth issue of his parents. They had six children in all: three sons—Mahadev, Seetaram and Keshav; and three daughters—Saroo, Rajoo and Rangoo. The frolics of the children enlivened the atmosphere of the household, and spread joy all around. From the moment of offering *pranam* to the family deity in the morning, up to the time of reciting "*Shubham karoti kalyanam*" after lighting the lamp in the evening, the children were constantly exposed to traditional samskars and cultural ways.[3]

For nearly a score of years, at the close of 19th century and the beginning of the 20th, there were repeated outbreaks of plague in the country. Out of the one lakh population of Nagpur, two to three hundred people died every day for weeks. Baliram Pant Hedgewar was ever ready to help the families of the deceased. On some days he had to visit the crematorium as many as twenty times, but he never shirked from this religious obligation. He had to take bath a number of times each day. He refused to vacate the house, though many other families had moved to escape the dreaded epidemic, and consequently died along with his wife.

As if to make their cup of misery full, with the passing away of the elders, Mahadev Shastri took to a loose and indisciplined life. And the younger brothers Seetaram and Keshav had to bear the brunt. The latter had to do all the household chores, from splitting firewood and storing water to cooking. On occasions they had to go without food. They had long since got used to moving about in torn clothes. And to add to it, they had to stand gratuitous abuse and beatings from their elder brother. Disgusted with all this, Seetaram Pant left the house and went away to Indore to pursue his Vedic studies. Keshav spent most of his time in friend's houses.

But Keshav was highly self-respecting. Even when he was tormented by extreme hunger, he never accepted food from his friends. He never begged for money. His interest in studies continued unabated. His name invariably appeared among the top five of the class. Calm and given to few words, he was of a very warm disposition.

In those days Dr. B.S. Moonje was the most outstanding public figure in Nagpur. He drew to himself youths of various persuasions, from constitutionalist Moderates to firebrand revolutionaries. Keshav, naturally, gravitated towards Dr. Moonje. Lokmanya Bal Gangadhar Tilak had just then launched his 'paisa fund' through his newspaper *Kesari*. Keshav worked ceaselessly in Nagpur for the success of that project. He went from house to house for fund raising. A true social worker neither asks for anybody's favour for himself nor is he ashamed to beg for a social cause. And this attitude, Keshav had imbibed since his student days.

Attending public meetings and listening to leaders speeches was Keshav's favourite pastime. He and his friends naturally nursed the ambition that they too would some day be able to deliver lectures for propagating patriotism. As a training-ground, they started a debating society and arranged periodical debating sessions. This activity received great encouragement from the group called the *Swadesha Bandhava* organized by Dr. Pandurangrao Khankhoji, who later became a famous revolutionary. To encourage consumer products of Indian manufacture, a store *Arya Bandhava Veethika* was opened by the *Swadesha Bandhava.* The brand was of 'five weapons'—which carried its own message. Keshav frequently worked there as a salesman. He also spread the message of Swadeshi among the friends.

After passing his matriculation examination in the middle of 1910, Keshavrao left for Calcutta, seven hundred miles away from Nagpur and a place totally unfamiliar to him. He carried a letter of introduction from Dr. Moonje. Soon after entering the College, Keshavrao began cultivating the acquaintance of students from various provinces. In his free time he visited the different hostels. He endeared himself to everybody by his warm and good humoured disposition. As in Nagpur, in Calcutta too Keshavrao continued his physical exercises and naturally consumed substantial quantities of milk. He thus possessed an impressive physique, chest and powerful shoulders. His indeed was a figure with 'muscles of iron and nerves of steel.'

Keshavrao passed the final examination and secured the L.M.S. degree in 1914. The convocation was held on 12th September 1914. In accordance with the requirement, he underwent practical training for the prescribed period and completed the course on 9th July 1915. He had on hand the offer of a lucrative job in Bangkok, but he refused. He had in fact no intention of taking up any job anywhere.

However, when the First World War broke out and the government began recruiting doctors to serve on the warfront, Keshavrao felt it to be a unique opportunity to gather the war experiences at first hand. He and a few of his close friends hastened to meet Suhravardi Khan and get their names enrolled in the army. However, since their names were already on another list, their applications proved to be mere dead letters.

Degrees from the national institutions were not being recognized by the government in those days, and were frowned upon by it. Sir Parde Lukis had brought a motion in this matter in the Central Legislature. That legislation just enabled the native graduates to carry on their profession, but the government would not consider as valid the certificates issued to patients by these doctors. The intention was to discourage the institutions which had defied the government and were being run on public support. In order to counter this partisan and arrogant attitude of government, Keshavrao started an unusual movement, with the co-operation of the new as well as old graduates.

Reports suddenly began to appear in the newspapers of Bengal and other provinces of meetings held in Calcutta and elsewhere to protest against the 'Bogus Medical Degrees Bill'. On the basis of these reports, editors of papers wrote scathing editorials. This naturally upset the government officials. But what puzzled them most was how the government espionage agencies were completely in the dark about these protest meetings.

The fact was that these meetings had never really taken place at all! Before launching the move, Hedgewar had met Ashutosh Mukherjee and explained the plan. Ashutosh fully supported it, and also emphasized the need for securing wide publicity in papers. Hedgewar accordingly met leading newspaper editors. Motilal Ghose, the reputed editor of Amrit Bazar Patrika promised his fullest co-operation for the movement.

The "protest meetings" in various parts of Calcutta were "held" by Hedgewar sitting in his own room. Reports of these "meetings" also emanated from his room. The papers promptly carried the reports. Prominent leaders were also mentioned as having participated in the meetings. Hedgewar had met these persons in advance and intimated the strategy to them. Thus, even with the best of efforts, government intelligence failed to get advance information about these "meetings". When they went and checked with the individuals who allegedly presided over or spoke at the meetings, they straightway corroborated the newspaper accounts. The meetings thus eluded the informers, who earned the chastisement of their senior officials.

Arrangements were also made to transmit the reports of Calcutta papers to papers in other provinces. They too castigated the policy of the government. At the height of the 'movement', however, Hedgewar actually did arrange a mammoth public meeting in Calcutta. The well known national leader of those days, Surendranath Banerjee presided over the meeting. The resolution adopted there voiced the public demand in no uncertain terms. The government was forced to yield. It agreed to accord recognition to all the medical graduates after holding a token examination.

The examination was held on 3rd November 1915. Many took advantage of it. Keshavrao was also permitted to take the examination. He, however, refused to appear, saying, "We have all studied in the National University. That University has examined us and awarded its degree. Where then is the question of government recognition ? It is precisely to avoid having to submit to alien dictation that we boycotted government educational institutions !"[4]

It is obvious that Hedgewar had no personal axe to grind in organizing the protest; he never had any desire to secure government's recognition to his own medical certificate; his sole intention was to expose the anti-people nature of the alien rule and arouse public indignation against it.

After completing the five-year course, Hedgewar returned to Nagpur in early 1916. Even prior to leaving for Calcutta he had determined not to become "Your most obedient servant" to the government. He returned to Nagpur with an even greater resolve about his future course of action. Many of his co-students, however, anxious to 'settle down in life', conformed to government requirements and started their professional careers. But as far as Hedgewar was concerned it was not for earning money or personal prestige that he had studied. When the government had launched prosecution against Aurobindo Ghose and he submitted his resignation from the principalship of the National University, he had said: "It was not to provide some book learning or means of livelihood to young men that we have started this educational institution. Our aim has been to prepare them to serve the Motherland and undergo suffering and sacrifice if need be at Her altar."[5] Hedgewar volunteered to fulfil that high expectation.

For sometime Hedgewar left the field of Active Resistance (Militant Nationalism) and adopted the path of Passive Resistance (*Satyagraha*). From a hard core revolutionary, he became a true Gandhian *Satyagrahi*. This was the second turning point of his life. The first was when he had transformed his love for religion into his love for the Motherland under the influence of Bal Shivaji's upright conduct in the court of Bijapore.

Now, he became a passive resister in place of an active resister. In other words, a non-violent Gandhi was born in him in place of the Chapekar Brother with a gun. And this was the second stage of growth of Hedgewar from a man of Militant Nationalism to a man of truth and non-violence of Gandhian concept.

He entered the field of active politics and became an active member of the All India National Congress in August 1919 and started taking part sincerely in all the activities of the organisation. Soon by virtue of his hard work, dynamic personality and unique ability in the field of organizational activities, he came to be recognized as one of the strongest pillars of the Congress in Nagpur.

As a responsible member of Central Province and Barar P.C.C., he toured the Central Provinces and the region of Vidarbha. He found the political situation to be highly confused. On the one hand was the massacre of hundreds of innocent people at Jallianwallah Bagh on April 13, 1919, which had shocked the Indian conscience beyond repair and set the young national blood on fire. On the other was the Non-violent Movement launched by Gandhiji on November 24, 1919 from the platform of the All India Khilafat Committee for the restoration of the Khalifa of Turkey, the movement which was wedded to peaceful and non-violent means, demanding strict discipline on the part of *Satyagrahis*.[6] The Militant School wanted aggressive action, the Moderate School, passive; and the two overlapped each other so much in the field of politics that it was hard to distinguish between the two. And the situation came to be confused further when Gandhiji called upon the Hindus to do and die for the cause of the Caliphate with their Muslim brethren. "If the Hindus wish to cultivate eternal friendship with Mussalmans", he said, "they must perish with them in the attempt to vindicate the honour of Islam".[7]

Hedgewar disapproved Gandhiji's policy of launching the Non-cooperation Movement with Khilafat as its main objective. The Khilafat agitation, aiming to restore the Caliphate in Turkey, Hedgewar argued, would only breed extra territorial loyalty and religious fanaticism among the Muslims of India. But he was not the man to stand aloof as a passive spectator during a national struggle merely because it did not come up to his expectations on all points. So, in spite of his reservations about the Khilafat agitation, he went to Amritsar to attend the Congress session which was to be held in the last week of December 1919. Hedgewar went to Jallianwallah Bagh first, saw the blood stained site, applied Tilak with the *mati* of the martyrs' land and felt that it was his blood that had spilt here, there and everywhere.

He attended the Congress session, took part in its deliberations, met the leaders of different hues. He saw that the Extremists were quite a live force under the leadership of Lokmanya Tilak, and that the situation in the Congress was fast reaching a flash point for a show of strength between the Moderates and the Extremists.

And Hedgewar, being an ardent lover of Tilak and instinctively inclined towards Militant Nationalism, honestly wished that Tilak should control the Congress platform. The next Congress session was to be held in Nagpur, the heart land of the Extremists. Hedgewar and his friends decided that the next session would be presided by Lokmanya Tilak, and for that purpose they plunged themselves headlong into preparations.

Preparations started in full steam. A reception committee was formed and the publicity campaign was launched. One report in the periodical *Maharashtra* read: "Accompanied by the leaders, the villagers elaborately decorated the villages and welcomed the guests. The enthusiasm of the peasant folk was unprecedented. The lectures by Moonje, Hedgewar, Ganpathrao Joshi and Babasaheb Deshpande cast a spell over the masses."[8]

While fervent preparations were going on, came the tragic news that Tilakji had passed away in Bombay on the night of July 31, 1920. For the Extremists it was a bolt from the blue, and especially for Hedgewar who was pinning high hopes on him. However, it was decided that Aurobindo Ghose should be persuaded to come from Pondicherry and preside over the Congress session. And for that purpose Moonje and Hedgewar went to Pondicherry. But their mission failed because Aurobindo refused to leave Pondicherry and rejoin agitational politics after renouncing it.[9]

And then came the historic Nagpur session of Congress held in the last week of December 1920. It was attended by over 3,000 members of the Reception Committee, nearly 15,000 delegates and seven to eight thousand spectators.[10] Hedgewar was made incharge of lodging and boarding and looking after the other conveniences of all the members. He worked hard, day and night, and worked so meticulously that it satisfied all the members. All were thoroughly impressed with his rare skill for managing the affairs. Also they were highly delighted to find a rare type of humility and service mindedness in a person of his age.

Hedgewar took part in the deliberations also. He and his friends drafted a resolution which spoke of 'Complete Independence'. They met Gandhiji and pleaded with him that Congress should endorse the resolution. But Gandhiji dismissed their suggestion by saying that the word 'Swaraj' incorporated the spirit of their resolution.[11]

Like a hurricane Hedgewar moved from place to place, lighting the torch of freedom in the hearts of people. And wherever he went, he created ripples. Men and women, young and old, gathered in thousands to listen to him, and big bonfires of foreign clothes would follow when his meetings ended.

This alerted the British government. Cyril James Irwin, the District Collector of Kotol, passed an order on February 23, 1921, imposing restrictions on his participation in public meetings, and lecturing and addressing gatherings of more than five persons. He ignored the order and continued to tour and lecture as before. His speeches at Kotol and Bharatwada were found "highly objectionable". He was sued on May 31, 1921 for violating the restrictive orders and spreading "sedition" against the government.[12]

The hearing began on June 14 in Judge Smely's court. Because of the biased attitude of the Judge, Hedgewar's lawyer found it difficult to carry on his defence. He, therefore, prepared to plead his case himself. He read out his written statement on August 5, 1921, as under:

1. It has been charged that my speeches have spread discontent, hatred and feeling of sedition towards the British Empire in the minds of Indians and sown seeds of enmity between Indians and Europeans. And I have been asked to explain. I consider it an affront to the dignity of my great country that a foreign government should subject a native Indian to inquiry and sit in judgement.

2. I do not recognise that there exists in India today any lawfully established government. It will be surprising if anybody should claim so. What obtains today is a regime of usurped authority and a repressive rule deriving power therefrom. The present laws and courts are but handmaids of this unauthorised regime. In any part of the world, it is only a government of the people constituted for the people that are entitled to administer law. All the other forms of rule are but ruses adopted by deceitful usurpers to loot helpless nations.

3. What I tried to do was to inspire in the hearts of my countrymen an attitude of reverential solicitude for their motherland which at the moment happens to be in a wretched condition. I tried to instill in the people the conviction that India belongs to Indians. If an Indian speaking for his country and spreading the nationalist feeling is regarded as committing sedition, if he cannot speak the truth without promoting hatred between Indians and Europeans, Europeans and those claiming to be the Indian government will do well to bear in mind that the day is not far off when foreigners will be forced to quit this country.

4. The government's version of my speech is neither accurate nor complete. Some stray notes and absurd sentences have been sloppily put together. But that does not bother me. In dealing with Britain and Europeans I have borne in mind only the basic principles that ought to govern the relationship between two countries. Whatever I have said has been with a view to asserting the birthright of my countrymen and the inevitability of securing our independence. I am prepared to stand by each word that I have uttered. Though I cannot say anything else concerning the charges against me, I am prepared to justify each word and letter of my speech and I declare that whatever I have said is lawful.

He further said, "India belongs to Indians". We therefore demand independence. This is the content of all my speeches. People have to be told how to secure independence, and also how to conduct themselves after securing it. Otherwise it is quite likely that our people may imitate the British in Independent India. The British, though they have their own territory, are not content with it. They are aggressing on other nations and governing them through repressive measures. But the very same British people are ready to shed blood when their own country's independence is threatened. The recent war bears testimony to it. We are therefore obliged to advise our people, 'Dear country-men, do not imitate the aggressive ways of the British. Secure independence by peaceful means; and be happy and content with your own country without hungering for other's territories.' In order to explain this idea I cannot avoid referring to current political issues. That the British have been carrying on their despotic rule in our beloved country is obvious to everyone. What law is there that gives one country the right to rule over another? I am asking you, the counsel for government, this simple and straight question. Can you answer it? Is it not against natural justice? If it is true that no country has a right to rule over another country, who gave the British the authority to trample the people of India under their feet? Do the British belong to this land? How then can they enslave us and declare that they own this country? Is it not the most blatant murder of justice, morality and *dharma*?

"We have no desire to dispossess Britain and rule over it. Just as the British in Britain and the Germans in Germany rule over themselves, we who own this country of India wish to rule over ourselves and carry on our own affairs. Our mind revolts at the thought of remaining the slaves of the British Empire and carrying that stigma for all time. We demand nothing short of Complete Independence. Till we achieve it we cannot be at peace. Is our desire to be free and independent in our own country against morality and law? I believe that law exists not to demolish morality but to enforce it. That ought to be the prime purpose of law."[13]

It was no surprise that the presiding judge exclaimed: "His defence is even more seditious than his original speech." In his judgment delivered on August 19, Justice Smely ordered Hedgewar to give an undertaking in writing that he would not deliver seditious speeches in future for one year and furnish bail of rupees three thousand.

Hedgewar's reaction was sharp and short:

"My conscience tells me that I am completely innocent. A policy of repression would only add fuel to the fire already raging because of government's vicious policies. I am convinced that the day is not far off for the foreign regime to reap the fruits of its sinful actions. I have faith in the justice of the Omnipresent God. I, therefore, refuse to comply with the order of the bail."[14]

Even as Hedgewar finished his reply, the judge pronounced one year's rigorous imprisonment for him. While going to the jail he said to his friends and well-wishers who had assembled there, "As you are aware, I have defended myself in this case of sedition against me. However, these days, there is an impression going round that arguing in one's defence is an act of treachery to the national movement. But I feel it is highly unwise to merely get crushed like a bug when a case is foisted upon us. It is our duty to expose to the whole world the wickedness of the foreign rulers. That would indeed be an act of patriotism. And not to defend ourselves; on the other hand, would be a suicidal policy. You may, if you so choose, refuse to defend yourself; but for God's sake don't consider those who disagree with you as being less patriotic. If in the course of our patriotic duty we are called upon to enter the prison or to be transported to the Andamans, or even face the gallows, we shall have to willingly do so. But let us not be under the illusion that jail-going is all in all, that it is the only path for achieving freedom. There are, in fact, so many fields of national service awaiting us outside the prison....I would be back amongst you after one year. Till then, of course, I will not be in touch with the national developments, but I am confident that by then the movement for Complete Independence will have gained added momentum. Now, it is no more possible to keep down Hindusthan under the heels of foreign domination. I offer my gratitude to you all and bid you good-bye.'[15]

Hedgewar was imprisoned in the Ajani jail, Nagpur, for one year. There he went through the jail schedule, including hard labour, ungrudgingly. By way of hard labour, he was given the job of polishing hand made paper. The crude chemical used for glazing paper frayed his hands. But he was not a man to complaint. On the other hand, he helped the jailor, Sir Jather, who was newly appointed, to understand the jail manual in detail and work out the situation peacefully whenever he was in trouble. About Hedgewar the jailer, Sir Jather, had noted in the Jail Diary: "Hedgewar had no interior motive of securing some favours or to come to some underhand arrangement. To his close friends Sir Jather used to say: "The courteous and winning manners and disciplined life of Hedgewar in jail had impressed us most. Even though we were government servants, we were so much drawn to Hedgewar by his amicable behaviour that after his release, whenever we went to city, our feet would automatically move in the direction of his house."[16]

He spent his free hours either spinning with the takli or reading the Shantiparva of the *Mahabharata* or the *Bhagwad Gita*. But the worsening condition of the country continued to hang heavy on his heart.

Hedgewar was released from jail on July 12, 1922. When he came out, he had put on an extra 25 pound. What sort of carefree life in jail would he have lived? Never shirking from hard labour, never breaking the jail discipline, never asking for any favour and still adding extra weight of 25 pound! This showed his attitude towards life in general.

And to honour him a grand meeting was arranged in Vicktarh Theatre, Nagpur, the same evening. But amidst all that grand festivity and praise showering exercise, he sat silent, quiet and composed, lost in himself like a philosopher. Hedgewar's reply was brief and poignant: "We have today to place before the country the highest and noblest of ideals. Any ideal short of 'Complete Independence' will take us nowhere. Even if death were to stare us in the face, we are not to shirk from our path. We have to keep the ultimate goal constantly burning in our mind and calmly carry on the fight".[17]

One more point which he clarified during his speech was regarding the concept of 'non-violence'. "Real non-violence," he said, "lies in the attitude of mind. At heart one should not harbour feelings of violence or hatred. One may outwardly carry out a certain act which appears to involve physical violence, but if it is done in a spirit of detachment and without any selfish motive or hatred attached to it, then the said act can no longer be termed violent. This is what Shri Krishna says in the Bhagwat Gita".[18]

Coming out of jail, he tried to absorb himself in different activities but failed to get himself attached to any of them for a long time. First, he busied himself in the activities of the *Anath Vidyalaya Griha* to save orphans from Christian missionaries and shelter revolutionaries on the run. Then he joined the Rifle Association and the Swadhyaya Mandal. He came on the Provincial Sports committee. He became Joint Secretary of Nagpur Provincial Congress Committee. He edited a daily newspaper called *Swatantrya*. And, above all, he toured the Central Provinces and the state of Maharashtra to spread the culture of 'Truth' and 'Non-violence.'

But all those activities failed to satisfy him. No doubt wherever he went, he was received enthusiastically by big gatherings. High profile processions were held in his honour. Very high sounding eulogies were sung in his praise. But amidst all those rejoicings and congratulations, he was found most of the time calm and reserved. Though outwardly he responded to the grand receptions, inwardly he was heavy with sadness at the worsening condition of the nation. All those honours, and felicitations showered on him, all those offices and chairs he held, all those eulogies sung in his honour did not sooth his troubled heart. The politician in him was slowly dying; the Yogi in him had begun taking roots. He was lost in himself most of the time in search of 'something' which could prove healthy for the failing health of the nation.

Practically, the last political meeting he attended, before going into self exile in search of that 'something', was on October 18, 1922. Before bidding good bye to the Congress culture, he also hinted in that meeting about the mode of life he was going to adopt for himself. This was going to be the third turning point of his life. The second was when he had left behind the life of a fire eating revolutionary and adopted the life of a true non-violent Gandhian *Satyagrahi*. The third was going to be a departure from political life of meetings and demonstrations to enter into a new life of silent *sadhna* and *tapasya* in search of that panacea which could cure all the ills of the Motherland. Hedgewar never left the field of politics but his approach to problems changed altogether.

REFERENCES

1. Hingle, G.S., *Hindutva Reawakened* (New Delhi: Vikas Publishing House Pvt. Ltd.), 1999, p. 1.
2. Seshadri, H.V. (ed.), *Dr. Hedgewar: The Epoch Making* (Bangalore: Sahitya Sindhu Prakashana), 2002, p. 2.
3. *Ibid.*, p. 3.
4. *Ibid.*, p. 29.
5. Das, Swaramala, *Sri Aurobindo: A Political Thinker* (New Delhi: Ajanta), 1993, pp. 133-153.
6. *The Collected Works of Mahatma Gandhi*, Vol. XVII (New Delhi: Publications Division, Ministry of Information & Broadcasting, Government of India), 1990, p. 312.

7. *Ibid.*, pp. 306-312.
8. Seshadri, H.V. (ed.), *Dr. Hedgewar: The Epoch Making, op. cit.*, p. 44.
9. Joshi, V.C. (ed.), *Sri Aurobindo: An Interpretation* (Delhi; Vikas Publishing House Pvt. Ltd.), 1973, p. 8.
10. Zaidi, A.M. (ed.), *The Encyclopadeia of Indian National Congress*, Vol. VII (New Delhi: S. Chand), 1979, p. 594.
11. *Ibid.*, pp. 591-696.
12. Seshadri, H.V. (ed.), *Dr. Hedgewar: The Epoch Making, op. cit.*, p. 49.
13. *Ibid.*, pp. 51-52.
14. Hingle, G.S., *Hindutva Reawakened, op. cit.*, p. 26.
15. Seshadri, H.V. (ed.), *Dr. Hedgewar: The Epoch Making, op. cit.*, p. 53.
16. *Ibid.*, p. 55.
17. Hingle, G.S., *Hindutva Reawakened, op. cit*, p. 29.
18. *Ibid.*

2

IDEOLOGY AND PERCEPTION

Hedgewar was from his early boyhood intensely devoted to the cause of nation and was in the thick of the various movements going on at that time for the liberation of our motherland. He was engrossed for several years in the revolutionary movement. He later on plunged into the agitations and organisational work of the Indian National Congress.

But soon he discovered that the underlying idea in all those movements was merely one of ousting the British. And that idea was equated with nationalism. For most of the leading men of those times, 'anti-Britishism' and 'nationalism' were inter-convertible terms. He felt dissatisfied with that superficial kind of thinking. He knew from a deep understanding of the history of the nation and also of other nations that the concept of 'nation' was a positive one and was not based on antagonism to anyone else. And he also knew that the slightest distortion in the initial concept of the national goal would, in spite of the best of intentions, ultimately lead to irretrievable catastrophic results.

Let us start with stating and analysing its concept as understood and practised in the modern world, as that would help us appreciate the subject in the present context.

What Modern Scholars Say

How did the concept of a 'nation' first manifest itself in the West? To begin with, various groups of people delimited themselves within some sort of natural territorial boundaries. The people residing in a particular territory developed a feeling that they were the sons of that soil, that they had their own way of life which they had to preserve and protect, and that their interests were not the same as those of other similar groups, and that, in short, they were a single, separate and distinct entity. They thus formed themselves into cohesive, indivisible groups. At various times and from various countries, leading men of thought gave expression to the idea of a 'nation' to define those groups. If we collect the various expressions and definitions put forth by them and bring out their essence, we arrive at certain simple conclusions.

The first requisite for a nation is a contiguous piece of land delimited as far as possible by natural boundaries to serve as the substratum on which the nation has to live, grow and prosper. Then the second requisite is, that the people living in a particular territory should have developed love and adoration for it as their motherland, as the place of their sustenance, security and prosperity. In short, they should feel that they are the children of that soil.

Then, people should not be just a mass of men, or just a juxtaposition of heterogeneous individuals. They should have evolved a definite way of life moulded by community of life-ideals, of culture, of feelings, sentiments, faith and traditions. If people thus become united in a coherent and well ordered society, having common traditions and aspirations, a common memory of the happy and unhappy experiences of their past life, common feelings of friendship and hostility, and all their interests intertwined in one identical whole then such people living as children of that particular territory may be termed a 'nation'.[1]

Hindu—The Child of Bharat

If one applies this definition acknowledged by all the learned men in the world, to India, one finds that this great country of ours, extending in the north from the Himalayas—with all its branches spreading north, south, east and west, and with the territories included in those great branches—right up to the Southern ocean inclusive of the islands, is one great natural unit. As the child of this soil, the well evolved society has been living here for thousands of years. This society has been known, especially in modern times, as the Hindu Society. For, it is the forefathers of the Hindu people that have set up standards and traditions of love and devotion for the motherland. They also prescribed various duties and rites with a view to keeping aglow in our mind a living and complete picture of our motherland and devotion to it as a Divine Entity. And again it is they who shed their blood in defence of its sançtity and integrity. That all this has been done by the Hindu people alone is a fact to which the history of thousands of years bears testimony. It means that only the Hindu has been living here as the child of this soil. All the great personalities right from hoary ages down to modern times have, by their life and example, confirmed the continuity of this sacred bond.

The common adoration for the motherland had made the people, in a way, related by blood to one another from the man in Kashi to the man in Kanyakumari and from a forest dweller to a city dweller. The various castes, the various ways of worshipping God, the various languages are all expressions of the one great homogeneous solid Hindu people—the children of this motherland.

Community of Culture

The Hindu people, have had an ideal, viz. to realise the Ultimate Truth. In keeping with that ideal one has a *dharma* which is incomparable for its breadth of vision and encompasses all grades and aspects of human life. One has a current of life, our *sanskriti* (culture) which instils sublime qualities of purity, character, fortitude and self-sacrifice in the individual, enabling him to attain that highest goal of human existence. That stamp of culture is manifest in the day-to-day life, sublimating the mundane into the transcendent. For example, it is common custom with the Indians to call every woman, even a little girl, by the word *ma*, i.e., mother. The same meaning is conveyed whatever other words we may use in our various dialects. That is, every woman, whatever her age or status in life, except a man's wife, is a manifestation of the mother to him. This is a special feature of our culture.

The famous instance of Shivaji who sent back honourably and laden with presents the beautiful daughter-in-law of the Muslim

Subhedar of Kalyan captured in war (though it appears exceptional in the eyes of the foreign, especially Muslim, historians) is a very ordinary instance symbolic of the sublime culture of this land. And this is shared even by the commoners here.

There was a revealing instance during the holocaust after the assassination of Gandhiji. An orgy of murder, loot and arson had been let loose on innocent people by anti-national and subversive elements. One such murderous mob attacked a house. Only the lady of the house was in. Out of terror she opened the door. On seeing her alone, the mob paused for a while. Notorious ruffians were leading the mob. But even then the ancient call of our culture *mtruvat paradareshu* (look upon women as mother) which was in their blood, made them treat her with utmost courtesy. They told her that they were out to burn and loot, but would arrange for her conduct to a place of safety. Two of them were even deputed for her escort! In spite of the past thousand-year-long corroding influence of foreigners and the all round immorality raging among our 'top class' at present, such slowing instances of character are still to be found in our common people. This gives us an idea of the distinctively pure and sublime culture which has been reared in our hearts. It has given us a complexion of our own.[2]

Also, down all these thousands of years, the country has produced a whole galaxy of seers and sages, savants and heroes who have led us in our march in search of God and attainment of material affluence and honour. They built up a social order to meet the needs of various kinds of individuals, developed economic systems for the proper production and distribution of wealth and formed political institutions for the protection and upkeep of the orderly evolution of society. As a result, during all these centuries of the existence here, we have had similar experiences of prosperity and adversity, of friendship and hostility, all our interests having been inextricably joined together.

Whenever, by mistake, the Indians fought with one another, imagining that the interests of a particular group were in opposition to those of another and developed separate and mutually hostile kingdoms, they lost the battle against the foreigners and all were reduced to abject slavery and misery. Same was the result when one section of the people was friendly to some foreign power while another was struggling against it. The case of Prithviraj Chouhan is in point. His enemy was Mohammed Ghori. Jayachand made friends with Mohammed Ghori. Here were two prominent persons of our country, one of them an enemy and the other a friend of the same invader. The result was that Prithviraj was defeated and killed. Jayachand also was destroyed. Mohammed Ghori became the king. And the whole train of disasters over the last eight hundred years, and which in the present times has culminated in the Partition, was set in motion. On the other hand, whenever we realised, even for a short period, that our interests were indivisible, that our friends and foes were alike to all of us, such immense power was generated in our national life that the power of the foreigner lay shattered at our feet. All the requisites for making a full-fledged nation are thus fulfilled in the life of this great Hindu People. Therefore, we say that in this land of ours, Bharat, the national life is of the Hindu People. In short, this is the Hindu Nation.

Touchstone of Nationality

When we say "This is the Hindu Nation", there are some who immediately come up with the question, "What about the Muslims and the Christians dwelling in this land? Are they not also born and bred here? How could they become aliens just because they have changed their faith?" But the crucial point is whether they remember that they are the children of this soil. What is the use of merely our remembering? That feeling, that memory, should be cherished by them. We are not so mean as to say that with a mere change in the method of worship, an individual ceases to be a son of the soil. We have no objection to God being called by any name whatever. We, Indians are Hindus to the core. That is why we have respect for all kiths and religious beliefs. He cannot be a Hindu at all who is intolerant of other faiths. But the question before us now is, what is the attitude of those people who have been converted to Islam or Christianity? They are born in this land, no doubt. But are they true to their salt? Are they grateful to this land which has brought them up? Do they feel that they are the children of this land and its tradition, and that to serve it is their great good fortune? Do they feel it a duty to serve her? No! Together with the change in their faith, gone is the spirit of love and devotion for the nation.

Nor does it end there. They have also developed a feeling of identification with the enemies of this land. They look to some foreign lands as their holy places. They call themselves 'Sheikhs' and 'Syeds'. Sheikhs and Syeds are certain clans in Arabia. How then did these people come to feel that they are their descendants? That is because they have cut off their ancestral national moorings of this land and mentally merged themselves with the aggressors. They still think that they have come here only to conquer and establish their kingdoms. So we see that it is not merely a case of change of faith, but a change even in national identity. What else is it, if not treason, to join the camp of the enemy leaving their mother nation in the lurch?

An eminent American Professor once asked M.S. Golwalkar the question, "Muslims and Christians are of this land alone. Why don't you consider them as of your own?" To that, he put him a counter-question: "Suppose one of our countrymen goes to America, settles there and wants to become an American citizen. However, he refuses to accept your Lincoln, Washington, Jefferson and others as his national heroes. Would you then call him a national of America? Tell me frankly." He said, "No."

Then he told him, "Why should not the same criterion be applied to this country also? How can you say that we should call such people nationals who, while living here, work against the honour and traditions of the country, and insult our national heroes and objects of national veneration?"[4]

Call to Shake-off Slavery

So, all that Hedgewar believed is that the Muslims and Christians here should give up their present foreign mental complexion and merge in the common stream of the national life.

Everybody knows that only a handful of Muslims came here as enemies and invaders. So also, only a few foreign Christian missionaries came here. Now the Muslims and Christians have enormously grown in number. They did not grow just by multiplication as in the case of fishes. They converted the local population. We can trace our ancestry to a common source, from where one portion was taken away from the Hindu fold and became Muslim and another became Christian. The rest could not be converted and they have remained as Hindus. Now, how did the converts leave their ancestral home? Was it out of their own sweet will and out of conviction of the superiority of those faiths? Well, history does not record a single notable instance of that sort.

On the contrary, history tells us that the reason was the fear of death or coercion or the various temptations of power, position, etc., or the desire to please the powers that be by adopting their ways and customs and finally even taking to their faiths. There was a lot of deception also. A piece of beef or a loaf used to be thrown into the water tank of a village and the villagers ignorant of what had happened, used to take the water as usual. On the next morning the missionary or the moulvi would come and declare that since they had used the polluted water they had all lost their religion and the only way left for them was to join his fold! In this way whole villages have been converted to Islam in the North and to Christianity on the West Coast. This is deception, pure and simple. Thus it was the mad zeal for increasing one's numbers for political domination. It was not propagation of religion, but a political strategy under its garb. The foreign invader not only subjugated them politically and culturally but ultimately converted them to his faith. That too is foreign domination. There are political, economic and cultural dominations and this is religious domination.

It is our duty to call these, our forlorn brothers, suffering under religious slavery for centuries, back to their ancestral home. As honest freedom loving men, let them overthrow all signs of slavery and domination and follow the ancestral ways of devotion and national life. All types of slavery are repugnant to our nature and should be given up. This is a call for all those brothers to take their original place in our national life. And let us all celebrate a great Diwali on the return of those "prodigal sons" of our society. The parable of the "prodigal son" is to be found in the Bible. A son parted from his family taking away his portion of wealth. He lived extravagantly and became a pauper. He returned to his place, but not having the courage to go home, stayed out of the precincts of the village. The old father who was on his way to his farm espied his son standing at a distance with his head hung in shame. The father immediately called him to his side, embraced him and exclaimed, "Oh, I am so happy that you have come back." He asked his other son to go home and arrange for a big feast. The other son was perplexed and asked his father, "Father, I have been here devoted to you all along. But you have never held a feast in my honour. This fellow has squandered away all his money and you intend to have a feast for him!" The father replied, "Look here, my boy, true, your brother had gone away for some time and I thought that I had lost a son. But now by the grace of God I have got back my son. Should we not love him as before and honour him? Should we not rejoice at his homecoming?"[5]

So also we shall rejoice and offer our love and respect when all those our brethren who have been wandering for so many centuries outside our house come back to our fold. There is no compulsion here. This is only a call and request to them to understand things properly and come back and identify themselves with their ancestral Hindu way of life in dress, customs, performing marriage ceremonies and funeral rites and such other things.

There are some people who claim that they have achieved unity of Hindus, Muslims, Christians and all others on the political and economic plane. But why limit the oneness only there? Why not make it more wide and more comprehensive so as to fuse them all in the Hindu way of life, in our *dharma* and take them back as lost brothers? To those who speak of unity on the political and economic plane, we say that we stand not only for political and economic unity but also for cultural and religious unity. We have thrown open our homes, our sanctuaries and temples, our age old culture and heritage. Undoubtedly, this is a broader outlook.

Much thought was devoted to the subject before the name 'Rashtriya Swayamsevak Sangh' was decided upon for the organisation. A number of persons close to the Sangh were also consulted. Naturally, many suggestions were put forth, and objections raised against each name. Even the present name was confronted with the objection that if the word 'Rashtriya' was adopted then its doors would have to be kept open to all other people in the country, i.e., Muslims, Christians, etc. So the name 'Hindu Swayamsevak Sangh' was suggested, as the organisation was restricted only to the Hindus.

Why 'Rashtriya' and not 'Hindu'?

Then why was the word 'Rashtriya' finally adopted in preference to the word 'Hindu'?

There is an instance which correctly denotes the underlying viewpoint. In Bombay there is a locality known as 'Hindu Colony'. Some persons of that place thought that the foreign word 'colony', should be changed. They came together to change the name, Hedgewar who was in Bombay at that time, was also invited. When he was asked for his opinion, he said, "More than the word 'colony', it is the word 'Hindu' that requires to be changed first." All those who had gathered, being very staunch Hindus, felt shocked. They asked, "How is it you, being a staunch Hindu, are opposed to the word 'Hindu'?" Hedgewar replied, 'I am not opposed to the word 'Hindu' as such. I am only opposed to its use in the present context which denotes that in our own country the Hindus have formed a colony! How can there be a colony of Hindus in Hindustan? If some Englishmen come, stay in a particular locality in Hindustan and call it a 'British Colony' it can be understood. So also it is, if you go to England, live in a particular locality and call it a 'Hindu Colony'. But a 'British Colony' in England, an 'American Colony' in the United States and a 'Hindu Colony' in our land Hindustan are all absurd. Either you accept that this is not your motherland and that you have come from somewhere outside and formed the colony here, or you change the name 'Hindu'". But they could not realise that there was something incongruous and absurd in having a 'Hindu Colony' in our own country. So the name continues even now.[6]

From this we can realise why the word 'Rashtriya' and not 'Hindu' was used in naming the organisation. Hedgewar used to say that in our land the word 'Rashtriya' naturally means 'Hindu' and therefore the word 'Hindu' need not be used. He would say,

"If we use the word 'Hindu', it will mean that we consider ourselves only as one of the innumerable communities in this land and that we do not realise our natural status as the nationals of this country."[7] But, it is a fact that there is in our country a grave misunderstanding about what 'nation' is, to which the Bombay incident is only a pointer.

The British Game

How did this misunderstanding creep in? During the last one thousand years, there had been many foreign aggressions on our land. The invaders such as the Greeks, Hunas and the Shakas who came previously were either defeated and driven out or, if some of them stayed over, were absorbed in the Hindu Society. But the invaders who came during the last ten or twelve centuries could not be driven out. They could not be absorbed either. They remained a separate entity and ruled as foreigners in this land. The last rulers, the British, were ruling from several thousand miles away. Being shrewd, they knew that perpetuation of their far flung empire was possible only by knocking out from the minds of the people the faith which gave them inspiration and strength to fight for freedom. What was that faith?

The British found that the average man here adores his motherland. The people here were blended together as an organic whole, though for the time being their unity was disrupted. A supreme sense of self-respect, love of freedom and devotion to their culture and traditions were ingrained in their blood. Knowing that such an indomitable spirit of nationalism would spell disaster to their empire, they planned systematically to undermine the faith of the people in their motherland, in their *dharma* and in their heritage.

With that end in view, the Englishman began training the Indians systematically in various ways. The first thing they taught was that this was one great 'continent' and not a country. They said that we are not one people and one nation. There were some aboriginals here staying for ages who had been driven away into the forests and hills. There were also other types of aboriginals called the *Dravidai*. *Aryans* came from the North and with their superior strength of arms conquered this land and dominated over the rest.

That is, we were told we had no motherland, that most of us had come from somewhere outside and therefore were equally strangers and foreigners to this country. We were also told we had no *dharma*, no philosophy, no morals worth the name, and that all our past life was just one of unrelieved darkness, that the coming of the English here was 'divine dispensation,' for we could learn sitting at their feet the first lessons in culture, religion and orderly social and political life.

The imperialist designs of Macaulay, the brain behind the system of English education, were trumpeted aloud: "We must at present do our best to form a class who may be interpreters between us and millions whom we govern, a class of persons, Indian in blood and colour, but English in taste, in opinions, in morals and in intellect." And again, in 1836, he wrote to his father, "No Hindu who has received an English education ever remains sincerely attached to his religion…If our plans of education are followed up, there will not be a single idolater among the respectable classes in Bengal 30 years hence."[8] Even to this day, the same insidious indoctrination of young minds is carried on in the Christian convents here that there is no real Saviour other than Jesus Christ, that the Hindu religion is only a mass of superstitions, that without taking to Christianity they would all be doomed to eternal damnation and so on.

In such an atmosphere it was natural that the English educated believed that all the centuries of our past life having been one of ignorance and internecine quarrels, we were to start afresh as a nation. In the early days of our freedom struggle against the Britisher, we find in the speeches of many top ranking leaders the expression "we are a nation in the making" which means, we are not yet formed into a nation. Even today many leaders say that we are a melting pot, that *we* are yet to give shape and content to our national life.

Another theory which caught their imagination was that since all those residing in this land were under the common danger of a foreign rule, that common danger resulted in forging them all into a new nationality. It is well known that a common danger can build up a nationality which already exists at least in a seed form but cannot create it *de novo*. It can only rouse and strengthen a nation but cannot create an entirely new one. It infuses into the nation a spirit of unity and sacrifice born out of the will to resist and defy the foreign aggressors. That is all the implication of the theory of common danger. And here it was sought to be applied to achieve new nationality altogether.

Absurdity of 'Territorial Concept'

They forgot that here already existed a full fledged ancient nation of the Hindus, and the various communities which were living in the country were either as guests, the Jews and Parsis, or as invaders, the Muslims and Christians. They never faced the question how all such heterogeneous groups could be called as children of the soil merely because, by an accident, they happened to reside in a common territory under the rule of a common enemy.

The theories of territorial nationalism and of common danger, which formed the basis of the concept of a nation, had deprived positive and inspiring content of the real Hindu Nationhood and made many of the 'freedom movements' virtually anti-British movements. Anti-Britishism was equated with patriotism and nationalism. This reactionary view has had disastrous effects upon the entire course of the freedom struggle, its leaders and the common people.

Thus these thoughts go on to delineate quietly and patiently the portrait of the best Indian society and pattern of values in all spheres of life and culture, philosophy, art and social order, that Hedgewar seeks to hold before the mirror of the national mind. To see the wood in spite of the bewildering number of trees that crowd the landscape, it is necessary to see in a bird's eye view the full design of the landscape after which details will fall into an illuminating order.

Pride in the cultural heritage of the past is the recurring theme of his philosophy. Chapter after chapter points to some aspect or other of the features that bear inspiring value for present emulation in social life and incarnation into individual character and personality. Enough is said to vindicate faith in the best values of inherited culture and its capacity to guidance even today.

1. *Philosophy—Darshanas*: Vedanta as the typical philosophy of Indians envisages a vision of the universe steeped in spirit fulfilling a life of eternal joy. Individual souls as sparks of such a supreme cosmic spirit have the divine destiny of realising universality and becoming godlike. This assures equality of spirit among all souls. What higher basis for democracy can we have than this? What greater stimulus to generosity and self-sacrificing participation in the life of society? The nation is the grouping next to family and clan and neighbourhood in the service of which we grow in moral stature.

All groups who share in this vision and discipline can become equal compatriots but none else. Traitors cannot be assimilated into the national fellowship until they change in spirit.

2. *Toleration* : Springing from understanding and sense of identity, this philosophy of immanence has produced the Indian doctrine of toleration of all points of view, religious views and ways of life. For all are expressions of the same cosmic spirit and can reach the same goal at their own pace and in their own way in accordance with their own symbols, provided they are sincere in their aspiration and effort. Errors will correct themselves in the climb.

Adhikara Bheda: Divergent status due to divergent qualification is formula that reconciles identity of goal and diversity of pathway.

3. *Social Order:* Such a view has evolved the outline of social order in India that enables members to pass normally from step to step in the ladder of life smoothly using the experience of ages and lifting their personalities to higher planes.

The individual career is distinguished into *brahmacharya, grihasthya, vanaprastha* and *sannyasa*. In the first, we have the student absorbing the cultural heritage, assimilating truth, science, philosophy, law and the arts. In the second, we have him fulfilling his duties as a member of society and as head of his household. The householder pays taxes, supports the state and does charity. The second half of life prepares the individual to detach himself from involvement in the detail of existence to contemplate the eternal. The values of life including the relations of the sexes in family and outside offer a pattern that is of universal value. It helps to make the best of life, with enjoyment, achievement and serenity in perfect equipoise.

The social order has another aspect, namely the duties and rights of different vocational groups viz., *swadharmas*. In addition to universal values like truth-speaking, non-stealing, non-injury, etc., Indian ethics recognises particular vocations—thinkers and men of worship i. e., free devotees of Truth and God or Nature or Reality; warriors and administrators, executives, etc.; traders and manufacturers and transporters, agriculturists and cattle tenders and breeders, etc. These classes emerge in the course of social differentiation in any progressive society. If their duties are laid down and social expectations are crystallised in society so that everyone will know what is expected of him, his rights as well as duties—a harmonious social order will result spontaneously. The present society will throw all careers open to talent on the basis of *swadharma* or duty springing from one's vocation and *karma* and *guna*, character and conduct.

No Class War: Another advantage of the Indian view of society is that it eschews class war. It postulates social harmony as a potentiality, if not as a full actual order of law and custom, observances and enforcements.

Dharma as determinant of the conduct of groups and individuals lays down a synthesis of the claims of conflicting parties. A study of the conflict impartially from the standpoint of the service involved, the function performed in society by the conflicting parties, will reveal the elements of the truth and justice in each party's claims. *Dharma* will come to mean an expression and crystallisation of the synthesis of interests and values revealed by the panchayat. This is the source of the long standing Indian faith in the panchayat way of settling disputes. It is democratic, scientific and moralistic at the same time. The destructive idea of class-war or irreconcilable antagonism between different functions, and classes is therefore successfully avoided in the Indian scheme.

The use of the word *dharma* does not preclude the inclusion of religious fellowships; different words may be used for the same meanings—like the Muslim and Christian. The substance of the thought and procedure and value is non-sectarian and human in the plainest sense.

Thus both from the standpoint of a philosophy of man and nature as well as from the viewpoint of social good, the Indian scheme is capable of affording a comprehensive social philosophy able to furnish a saving pattern for modern complex life—a social philosophy superior to current violent and muddy theories of Marxism and Freudianism.

Indian thought too has sound notions of the role of the State as the agent of common good—*raja dharma* which now may be designated *rajya niti* or *rajya dharma*. The State should be above partial interests and should regulate all activities according to *dharma*. The State is not a class or agent of upper class, according to Indian *shastras* or political and social science. Nor it is an exploiting agency. It is an agent of morality or *dharma*. It precludes socialism in the sense of adding economic to political power. The State is not a trader or manufacturer but is entitled to regulate all vocations. Thus Indian ideas help us to think through current conflicts and social theories and arrive at solutions more satisfactory because they are more truly based on wide experience and matured reflection.

No doubt we may surpass the past but only by thinking through them to new conclusions, not by ignoring them and starting *de novo*.

Also, it is a waste of effort if we are to start from scratch when great ideas are at hand for the asking. It is like each scientist ignoring all previous discoveries and starting from scratch from a false notion of originality!

It is not national chauvinism but sober common sense to build on the foundations already well and truly laid.

REFERENCES

1. Golwalkar, M.S., *Bunch of Thoughts, op. cit.*
2. *Ibid.*
3. *Ibid.*
4. *Ibid.*
5. *Ibid.*
6. *Ibid.*
7. *Ibid.*
8. *Ibid.*

3

NATIONAL RENAISSANCE

There has been a massive upsurge of community consciousness and identity politics all over the world in recent years. As the pace of modernization accelerates, the individuals are increasingly seeking to escape from individualism towards a more community based identity.[1] Every major philosophy of the present time considers individual alienation and cultural dislocation as important problems of man and society.

Community is a symbolic construction; it is a system of values, norms, moral codes, which provides a sense of identity to its members within a bounded whole. This concept thus carries both the notions of similarity and difference; the members of a group not only share certain commonalities with each other, they also possess certain characteristics which distinguish them from the other groups in a significant way. Individual identity is not just one's relationship with other members of the community; it is essentially a matter of being and it is this consciousness of belongingness to a particular Community that determines his identity.

Therefore, identity formation, generally a cultural construct, is a process of inclusion and exclusion of values and symbols defining 'we' and 'others'. In the process, boundaries between the two take shape whose forms keep changing from time-to-time. Relationships between 'we' and 'others' are not necessarily always conflicting or competitive; commonality and overlapping between the two in various social spheres often blur the boundaries. Further, the ascriptive identities, be they ethnic or communal, are products of changing historical contexts, and are not culturally fixed. However, this does not suggest that collective identities are undeveloped without clear boundaries. They are neither fuzzy nor determinate though at any given point of time they may appear to have been always there in exactly the same form.[2]

A community's awareness of its identity crystallizes when it finds a whetstone in another community, which is different from it. Cultural identity takes a political form and differentiation between 'we' and 'they' gets sharper and more concrete. Cultural identity is transformed into communal identity and community consciousness becomes communal consciousness. Now it is not just a question of one's belongingness to a community but also the intensity of this belongingness. In a 'communalized' situation, the intensity of communal consciousness is related to one's perception about 'others' who are considered enemies. The campaign against the 'other' is carried out through hostile propaganda, which completely ignores the commonality between 'us' and 'them' and only magnifies the differences. In the process, the meanings of 'we' and 'they' transcend the cultural mould and prejudices become pronounced. All these changes lead to communal violence.

Violence becomes inevitable when the individual identity of a substantial number of population gets submerged in the communal identity, championing love for 'us' and the hatred against the 'other' who is 'homogenized, depersonalized, and increasingly dehumanized. Rumors and rival mobilizations fuel a persecution mania.[3] The slightest spark leads to violence. This communal violence, in turn, further reinforces communal identity.

The origins of the nationalist movement in nineteenth century India can be traced to the expansion of Western education by means of the English language during the early part of the century. The English educated elites were concentrated in Bengal, Bombay and Madras, three coastal presidencies where greater opportunities existed for the new education and for vocations based on it—law, medicine, journalism, education, banking, modern trading and manufacturing. Those attracted to the new education came primarily from caste Hindu groups in all three presidencies. The proponents of social, political and religious reform among Hindus were drawn from this English educated class.

Until very late in the nineteenth century, most politically articulate Indians were willing to collaborate with the colonial administration. Political action took the form of petitions for greater Indian participation in the bureaucracy and in the legislative councils, protection of indigenous industry, reform of judicial administration and legal procedure to bring them into line with the practices established in England, and reforms of the Hindu social order that would make it more compatible with western norms.

Modernists and Revivalists

A shift from collaboration to criticism began in the latter part of the nineteenth century. This approach required new techniques of public protest. As older social, economic and psychological commitments were eroded among those most exposed to the change, newly acquired Western norms were employed to judge society. Two broad movements emerged among Hindus seeking to define their national identity, and we shall refer to them as the modernists and revivalists. The former adopted models of social and political change based upon Western patterns, and the latter looked to Hindu antiquity. Both groups found themselves wanting when judged by the new norms. Within both groups a broad range of ideological perspectives was present. Revivalism included those who wanted to preserve the traditional social order as well as those who sought to reform Hindu society as a way of strengthening Hindu solidarity.

The Hindu revivalists sought to recover fundamental truth which they felt had been lost since the era of its original revelation. They argued that the loss of national consciousness had created conditions conducive to foreign domination. By appealing to an idealized past, the revivalists reminded the Indian public of the suffering and degradation experienced under alien rule. The call for independence was a logical corollary, for the degraded present could only be overcome by eliminating the foreign intruders who had disrupted the original blissful society. Muslim rulers and the British were identified as sources of that disruption and many revivalist spokesmen sought to place limits on their political power and on their cultural influence. Consistent with the traditional Indian concept of knowledge, changes in Hindu society were justified by the proposition that the changes were not new at all, but were in fact a revival of older, purer form of Hindu culture that had degenerated during foreign rule.

The revivalists, unlike their modernist counterparts, spoke to an almost exclusively Hindu audience. The style of the revivalists was more aggressive and tended to reflect a *kshatriya* (warrior) world view. The identification with the martial tradition of Hinduism enhanced their self-esteem by the same standards used by many British to evaluate the worth of Indians. Research of Western orientalists confirmed the faith of the revivalists in the achievements of Hindu rule and civilization. The romantic nationalism sweeping Europe at the end of the nineteenth century provided additional intellectual justification for reconstructing the present by reaching back to the classical roots of the civilization.

Revivalism was expressed in new organizations which adopted to blend religious with socio-economic values to a revived sense of community and ultimately to espouse nationalism. Hindu society had degenerated, according to many early spokesmen of revivalism, because Hindus had not observed *dharma* (a code of conduct for various social categories, situations and stages of life). Many revivalists argued that India could not regenerate itself unless *dharma* were properly observed. Aurobindo Ghosh, maintained that "All great awakenings in India, all her periods of mightiest and most varied vigour have drawn their vitality from the fountain head of some deep religious awakening."[4]

The causative relationship between religious observance and socio-political affects represented a general commitment to Hindu metaphysics. Major schools of Indian thought offered a moral competition of existence based on the view that behaviour and natural phenomena are guided by inviolable laws. A good society can exist only when it is rooted on correct principles of *dharma.* The state's obligation in the classical Hindu texts was to protect the established order so that individuals could perform the duties prescribed by caste and stage of life.

To pursue and legitimize the political task of national integration and self-government, the revivalists selectively drew from a vast mosaic of often conflicting values. They redefined traditional concepts to justify public protest and to rationalize a restructuring of Hindu society. A central concern was to redefine the linked concepts of *dharma-karma*. Rules of *dharma* maintained social integration at the local level of Indian life and were enforced by a consensus of the elders of the local castes. The rules encompassed an individual's moral and religious life, and regulated social, political, and economic responsibilities. *Karma* refers to the willed actions of an individual. These actions have a good or bad effect which determines the quality of life in succeeding incarnations. Hindu practice tended to specify *dharma* according to the endogamus caste into which a person was born, though many of the Hindu sacred texts do not make this connection. A person was required to observe *dharma* unselfishly and without interest in the results of the action. At a more advanced stage on the path to *moksha* (release from the bonds of existence), the devotee subjectively seeks "realization" of the divinity. These concepts strengthened an identity with the small groups to which an individual belonged by birth and made collective identifications of a larger sort more difficult. According to P. T. Raju, the cumulative effect of Hindu thought is a negative attitude towards the material world.[5] Rajni Kothari proposes that the strains produced by national defeat and humiliation further encouraged other-worldliness, fatalism, and pessimism.[6]

Many revivalists, in their reinterpretation of *dharma-karma*, attempted to substitute the larger Hindu "nation" for primary group identification and to, create a life affirming orientation. Like millennial movements in Europe, the political character of Indian revivalism derived much of its strength from the importance of preparing for salvation.

Hindu Identity in pre-Modern India

To understand the Hedgewar arguments on Hindu identity and Hindu *Rashtra*, the term 'Hindu' has to be understood in a proper perspective. Probing the etymology of 'Hindu', one surprisingly does not find this word in Sanskrit, the language in which most of the texts of Hinduism have been written. Even the *Vedas*, the

Upanishads and *Bhagavad Gita* are silent on this word. Then how did this word originate? The term, it is argued, emerged as a geographical nomenclature and implied a geographical identity to the people living across the Sindhu or Indus river.

Then how does one explain 'Hinduism'? Ever since the term has been in vogue, scholars have been looking for an acceptable definition of it. However, the religious diversity of India has made it really difficult to find one. Did the people of this geographical area really identify themselves with an inclusive Hinduism? It is observed that multiple communities based on various identities, rather than some sort of inclusive Hinduism, existed in pre-modern India. Hinduism was a 'mosaic of distinct cults, deities, and ideas' adjusting to and distancing themselves from each other. The term 'Hindu' which emerged as a geographical nomenclature gradually came to be used not only for those who were living across the Indus river but also for those who professed a religion other than Islam or Christianity. This all inclusive term was doubtless a new and bewildering feature for the multiple sects and castes who generally saw themselves as separate entities'. Since Hinduism was a cluster of distinctive sects and cults, the people identified themselves with region, language and caste rather than with an inclusive Hinduism. Unless by 'Hindu' one means nothing more, nor less, than "Indian" (something native to, or found within the continent of India), there has never been any such thing as a single "Hinduism" or "Hindu Community" for all of India. Furthermore, there has never been any one religion nor even one system of religions to which the term 'Hindu' can be accurately applied.

It may be true that self-contained religious constellations existed in pre-modern India, which identified themselves with much more immediate and obvious domains of religious commitment rather than with the wider field of 'the Hindu tradition'. Yet the manner, in which the representatives of these theistic formations relate and refer to one another, indicates the presence of this wider field. Thus, it represents a sense of identity, a sense of coherence in a shared context and of inclusion in a common framework and horizon, and it refers us to some fundamental implications of the elusive reality of "Hinduism". It may be further argued that just because the term

'Hindu' was first used only in Arab texts does not make the application of the term Hinduism to earlier periods invalid. Despite the existence of a wide range of beliefs, from animism to atheism, the term Hinduism could still be applicable to the belief system of pre-modern India as one could see some of its features in the Hinduism of the later period.

Modern Indian Search for a Definition of 'Hindu'

At the turn of the twentieth century when community politics was in the ascendant, Hindu leaders began searching for a definition of the term 'Hindu', which would justify a separate Hindu identity. Different definitions were attempted, but each contradicted and turned down. Some defined Hindus as the people who burn their dead. Even this definition, based on the least common denominator, was found faulty, as a *sanyasi's* dead body is never burnt. Bal Gangadhar Tilak tried to end this uncertainty by proposing something different: A Hindu is one who believes that the *Vedas* contain self-evident and axiomatic truths. But this definition was also found to be incomprehensive and defective in as much as it excluded the Jains and Sikhs from the fold of Hinduism.[7] Thus, to the Hindu leaders, all such attempted definitions were partial and incomprehensive.

V.D. Savarkar's definition of 'Hindu' in his work *Hindutva* (1923) was a departure from the earlier Hindu leaders. He provided a clear and precise definition, which apparently resolved the dilemma of the Hindu leaders once and for all. Savarkar defined Hindu as 'a person who regards the land of Bharatvarsha from Indus to the Seas as his Fatherland as well as his Holyland that is the cradle land of his religion'.[8] This definition got applause and acceptance from the Hindu leaders and was heralded as 'a message given at the dawn of a new age'.

Hedgewar subscribed to Savarkar's definition of 'Hindu' and founded the RSS to propound India as a *Hindu Rashtra*. However, his successor, M.S. Golwalkar, who consolidated and amplified the ideology of *Hindu Rashtra*, seemed to have reservation on Savarkar's formulation. As he explains: 'Of course, many attempts at definition have been made from time-to-time, but all such definitions have proved to be incomplete. They do not express the whole truth and it is natural in the case of a people who have been growing and evolving for the last so many scores of centuries.[9] Thus Golwalkar did not accept the finality of Savarkar's definition of Hindu by projecting the Hindus as a 'growing and evolving' people.

Golwalkar admits the difficulty in defining the term 'Hindu'. For him, it is a paradox that though the Sun and the Moon could be defined easily, it seems impossible to define their creator, God. But does it imply that God does not exist, questions Golwalkar. The origin of the Hindus is unknown to the historians: 'In a way we are *anadi*, without a beginning.'[10]

Hedgewar admits that though *Hindu Rashtra* is the core of his ideology, he did not create this concept; he has only inherited this legacy from various Hindu thinkers of the nineteenth century, such as Vivekananda, Bankim, Aurobindo, Dayananda, Tilak, Savarkar and others. The nineteenth century was predominantly a period of cultural protest and adjustment in India. This produced new forms of community consciousness, particularly in three regions: Bengal, Punjab and Maharashtra. Hindu tradition was not immune to this development. A number of socio-religious reform movements like Brahmo Samaj and Arya Samaj emerged mainly 'as a reaction to the threat of Western domination and especially to a two fold cultural challenge, the utilitarian reformism and Christian proselytism'.[11]

These socio-religious reform movements attempted to defend, redefine and recreate Hinduism on the model of Christianity. They sought for the equivalent of a monotheistic God, a book, a prophet or a founder and congregational worship with an institutional organization supporting it. The ideology of these movements was of 'strategic syncretism': syncretism because there is a strong intention to reform one's society through the assimilation of Western values consistent with the Hindu cultural equilibrium, and strategic syncretism since the equilibrium in question remains the prime concern.

The nineteenth century 'cultural counter offensive' against the colonial rule has great variety and range, from emancipatory to divisive. It was a conscious effort to overcome the marks of subordination and humiliation inflicted by colonial rule. However, at times, this feeling of subordination extended back before the beginning of colonial rule, into the period of Muslim rule as well. For example, Swami Dayananda tempered his Hindu reformist zeal with a growing emphasis on Hindu unity, foregrounding cow protection and the replacement of Urdu by Hindi in the Devanagari script. With Dayananda 'rationalism gave way to faith, equivalence to superiority, and tolerance to militancy.' Thus, as nationalist thought took shape, the Muslims increasingly displaced the British in the category termed 'foreigner'.

Tilak and the Hindu Nation

Bal Gangadhar Tilak needs a special discussion for the following reasons: first, he not only attempted to draw the outline of a Hindu nation, he also actively and effectively mobilized the Hindus to crystallize Hindu identity; secondly, Hedgewar and his political guru Moonje were ardent followers of Tilak. The ideology of the RSS undoubtedly took its cue from Tilak.

In his writings *The Orion* and *The Arctic Home* in the *Vedas*, Tilak attempted to prove that his Aryan ancestors had actually planted the seed of civilization. Tilak's thesis have strong scholarly foundation; his conclusion generated a sense of psychological superiority among the Hindus. 'In a land where antiquity was generally equated with superiority, few contributions to the inflation of the national ego were of greater significance.'[12]

Tilak had developed his thinking in terms of a Hindu nation: 'The common factor in Indian society is the feeling of *Hindutva* (Hinduness).' When he talked of Indian society, he meant Hindu society. He considered the Hindus of different regions as one, all of them believing in a Hindu *dharma*. He observed, that there may be different doctrines in the Hindu *dharma*, but certain principles can be found in common, and because of this alone a sort of feeling that we belong to one religion has remained among people speaking different languages in such a vast country. And this feeling of oneness among the Hindus was further strengthened by different institutions of the Hindu religion like temples and places of pilgrimage, he analysed.

Tilak believed that religion was a vital element in nationality and argued that India was a self-contained nation during Vedic times. He observed: 'A Hindu of this place is as much as a Hindu as the one from Madras or Bombay . . . the inner sentiments which move you all are the same. The study of *Gita*, *Ramayana* and *Mahabharat* produce the same ideas throughout the country. Are not these common allegiance to the *Vedas*, *Gita* and *Ramayana* our common heritage?'[13] Tilak believed that the divergent sects of India could converge to form a 'mighty Hindu nation' if they would only follow the original principles of the Hindu tradition as set forth in such texts as the *Ramayana* and the *Bhagavad Gita*. And this convergence should be the goal of all Hindus. The views of Hedgewar very much reflect these ideas.

Tilak not only recognized *Hindutva*, he made an innovative attempt to mobilize the Hindus on the basis of this common denominator. During the 1893 Bombay communal riots he wrote that though the Hindus did not of their own accord attack others, the Marathas were not cowards to suffer meekly without retaliation. He emphasized that the only way for continued goodwill amongst the two communities was the consciousness of each other's strength.

The 1893 Bombay riots certainly soured Hindu-Muslim relations. Hindu leaders of Poona decided to boycott the Muharam festival, which had been celebrated jointly by both the communities till then. In 1893 a separate Hindu festival started. From being a purely private religious function, the Ganapati celebration became a public festival and a manifestation of Indian nationalism. This festival was also designed to wean away the Hindu lower classes from the ritual of another religion by promising them an equally attractive alternative.

The Ganapati festival was followed by a Shivaji one. Shivaji was another symbolic choice who made great sacrifice for the cause of Indian nationalism. If the Ganapati festival imparted to Hinduism a congregational character, the Shivaji festival went a step further by deifying the father of the Maratha nation, 'and thereby sanctifying its political history, thus providing Indian nationalism with a precept from its own past. Hedgewar has certainly borrowed from Tilak the technique of militant mobilization to assert and consolidate Hindu identity. It also emphasizes, like Tilak, the importance of demonstrating one's strength in a charged communal atmosphere.

Tilak observed that the central thrust of the *Gita* was *karma-yoga* (philosophy of activism) rather than *sanyasi's* (renunciation) and *loka-samgraha* (the highest good for all people) was the best end of such action. Therefore, his political activism became a religious activity, nay obligation (*dharma*), for all Hindus. Universal salvation was to be achieved by striving for the highest good for the people (*loka-samgraha*). In effect, Tilak 'made *Loka-samgraha* the equivalent of *Moksadharma* in this world (society)'. Hedgewar subscribes to Tilak's understanding of *karma-yoga* and *loka-samgraha* and the *swayamsevaks* are advised to read the *Gita* regularly.

V.D. Savarkar: Hindutva and Hindu *Rashtra*

Savarkar's definition of Hindu was a decisive step in Hindu politics. He defines 'Hindu' as a person who regards Hindustan as his *pitrubhoomi* (fatherland) and *punyabhoomi* (holyland). In Savarkar's *Hindutva,* two catch phrases were stressed: Hindu *Rashtra* (Hindu Nation) and *Hindutva* (Hinduness). He clearly distinguished *Hindutva* from Hinduism. To him, *Hindutva* is not a word but a history, not only the spiritual or religious history of the people but a history in full. 'Hinduism is only a derivative, a fraction, a part of *Hindutva*.'[14] Savarkar was clear: the Hindus have to profess Hindutva rather than Hinduism as their first defining characteristics.

Savarkar listed three criteria for confirming the *Hindutva* or Hinduness of a person—geographical, racial and cultural—common *Rashtra*, a common *Jati*, and a common *Sanskriti*. The first two essentials of *Hindutva Rashtra* and *Jati* are clearly denoted and connoted by the word *Pitrubhu*; while the third essential of *Sanskriti* is pre-eminently implied by the word *Punyabhu*, as it is precisely *Sanskriti* including *sanskars*, i.e. rites and rituals, ceremonies and sacraments, that makes a land 'a Holyland.' In essence, this emphasizes that a Hindu is one who follows a religion which had originated in India. Savarkar's idea of Hinduness, then, explicitly differentiates the Hindu Nation from the orthodox Hindu religion that acknowledges the authority of the *Vedas*. In this sense Sikhs, Buddhists, Jains, etc. were all Hindus. The logic of Savarkar is that all these originated in Bharatvarsha.

In his discourse, no particular Hindu symbol or rite is indispensable. However, these could be taken up in a pragmatic manner and resanctified, since *Hindutva* involves a 'common culture', a 'common civilization', and a 'common history' and so Diwali, Rakhi, and Hindu pilgrimages are discovered to be essential 'from a national and racial point of view'. Particularly significant in today's context is Savarkar's discussion of Ram's symbolic significance: 'some of us worship Ram as an incarnation, some admire him as a hero and warrior, all love him as the most illustrious representative monarch of our race'.[15]

Bankim and Aurobindo elevated nationalism to the status of a religion. For them 'Hindu' was basically a religious rather than a cultural or racial unit. Savarkar viewed nationalism and nationality not as a religious or a theological category. For Savarkar the term signified the cultural and the racial rather than the religious. In his discourse, 'religion if it entered at all was important only for its cultural connotations.' The modern Western idea of a nation, as Savarkar explained, could not be applied to India as it ignores the ancient glory of the Hindu people, the indigenous and numerically dominant population. Those people whose culture evolved and developed in greater India, from the Himalayas to the Southern seas, by some accounts from Iran to Singapore, constituted the Hindu Nation, according to Savarkar.

Though Savarkar did not reject the territorial concept of nationhood, he modified it substantially. He asserted that in the formation of nations—religious, racial, cultural and historical affinities—count immensely more than their territorial unity. Savarkar's position was that 'in Hindustan the Hindus were a nation, and other people were communities and numerically, therefore, minorities'.[16]

Hedgewar agreed on Savarkar's formulation of *Hindutva* and founded the RSS on the conviction that India is a *Hindu Rashtra*. It needs to be kept in mind that the founder of the RSS was a competent organizer rather than a brilliant ideologue. That is why he did not leave many ideological writings on the *Hindu Rashtra*. It was left to his successor M.S. Golwalkar to fill this gap. Golwalkar came out with a clear exposition of the ideology of *Hindu Rashtra* in his Monumental work, *We or Our Nationhood Defined* (1939). This work was later replaced by *Bunch of Thoughts* (1966). Another ideologue, who further enriched the concept of *Hindu Rashtra*, was Deendayal Upadhyaya.

Secularism

The term 'secularism' as used in the Indian context has two contradictory meanings: one is *dharmanirapekshata* and the other is *sarva dharma sama bhava*. While the former means indifference to religion and upholds the separation of religion and state, the later advocates an equal respect for all religions based on the principle of accommodation. Thus the debate on secularism oscillates between the two extremes of separating religion from politics and accommodating different religions through the intervention and assumed neutrality of the state. Yet there is an implicit connection between the two strands despite being juxtaposed against each other.

At one extreme, there are some secularists who advocate the need to keep a distance from religious matters, both at personal and political levels. One dominant ideological trend among them represents an unadulterated secularism, which completely dismisses the complexities of religion as obscurantist or reduces it to the level of superstition, a mere remnant of pre-modern states of oppression. At the other extreme, there are individuals and groups, who, despite being secular in their everyday lives, may be inclined to a certain idea of God or divinity. Though they draw a line between religion and politics, myth and history, they still believe that this dichotomy cannot be assumed on purely rationalist grounds.

RSS Critique of 'Pseudo-secularism'

The RSS and other Hindu nationalists have been questioning the genuineness of secularism pursued since Independence. They basically share three important objections raised by other critics with regard to the doctrine of secularism and to the State guided by it. First, that the concept is alien to Indian culture by virtue of its profoundly Christian/Western character. Secondly, the concept is absolutely insensitive to religious people. By propagating religion as a matter of private preference, it breaks the link between religion and community and deprives people of their sense of identity. Thirdly, though a secular state claims to be neutral, in reality it favours either the non-believer or the minority community.

The RSS argues that the secular state in India is far from neutral. In its view, the state, through various legislations, has hindered the growth of a Hindu identity. While the Hindus have been compelled to view themselves primarily as non-religious individuals, the Muslims are often encouraged to conceive their identity purely in terms of their religion. While the state officially professes neutrality, it is both antireligious and pro Muslim. Some *Sangh Parivar* members like the VHP even claim that the Indian state is anti-Hindu.

The RSS does not attack the term 'secular'. It has rather added the concept of 'positive secularism' to the existing ideological debate on secularism and branded all other practitioners of secularism as 'pseudo-secularists', conceding thereby its approval of the ideal as such of the secular state. But its secularism has a distinct feature: it is against any type of 'minority appeasement', particularly Muslim appeasement.

Features of 'Positive Secularism'

The RSS calls its own brand of secularism as true or positive secularism. Its salient features are: First, the RSS consistently clarifies that the *Hindu Rashtra* does not advocate a theocratic state. This could be believed, as the RSS is aware of its impracticability. Therefore it would rather prefer to have a state in which Hindus would enjoy a hegemonic role. It is very unlikely that in a *Hindu Rashtra* the laws of the state would conform to this or that *samhit*. In this matter the position of the RSS can be said to have undergone a considerable shift in position from that of the Hindu Mahasabha at the time of the debate over the Hindu Code Bill (1955). Its position is also quite different from that of most contemporary Islamic fundamentalist movements, which explicitly reject the theoretical separation of state and religion as Western and unIslamic.

The second corollary feature of this secularism is the usefulness of the modern state in the establishment of a uniform and homogenized national culture. The majoritarianism of the Hindu Right 'is perfectly at peace with the institutions and procedures of the Western or modern State'. "In fact the RSS understands that an interventionist modernizing State could be of immense help in obliterating the presence of religious or ethnic particularisms from the domains of law or public life and to provide a homogenized content to the notion of citizenship. In this role, the Hindu Right in fact seeks to project itself as a principled modernist critic of Islamic or Sikh fundamentalism and to accuse the 'pseudo-secularists' of preaching tolerance for religious obscurantism and bigotry.

The positive secularism of the RSS also swears by *sarva dharma sama bhava*. However, this interpretation is quite unlike the interpretation provided by Indian secularism. For the RSS, equal respect for all religions implies that all religions must be treated equally, that is, treated the same in law. Accordingly, any law or policies that provide special treatment for minorities are opposed, as being 'pseudo-secular' and appeasing the minorities. Article 30 of the Constitution, which allows minorities to run their own educational institutions, and Article 370 which grants special status to Jammu & Kashmir, are thus opposed as these provisions are nothing but minority appeasement rather than secularism. In consonance with its interpretation of equal treatment it demands a Uniform Civil Code which would treat all religious communities in the same manner.

Reasserting *Hindu Rashtra* and *Hindutva*

The present ideological writings of the RSS, besides denouncing the Nehruvian concept of a nation, also attempt to convince about the positive features of the *Hindu Rashtra* and *Hindutva*. Hindus are a nation—this is the present thrust of all ideological writings of the RSS. These writings passionately plead that the Hindus are just not a religious community, but a nation; Hindu *Rashtra* is not a political concept. Nation (*Rashtra*) and State (*Rajya*) are two different concepts and should not be mixed up. State is purely a political concept. Nation (*Rashtra*) is eternal while the State is transitory. If the State is the body, the nation is its soul (*Atma*). In Hindu philosophy the *Atma* is eternal and only the bodies and their forms are changed. By equating the *Rashtra* with the concept of *Atma*, the ideologues of the RSS plead that the Hindu Nation has been existing since time immemorial and it shall continue to exist forever.

Thus, the RSS strongly denies that the concept of *Hindu Rashtra* is not religious as it does not aim to establish a theocratic or a religious Hindu State. To substantiate this point, the RSS leaders speak of Hindu catholicity by quoting from '*Asya Vaniasya*' hymn of the *Rig Veda* quite frequently: *Ekam sad viprah bahudha vadanti* (Truth is one, the wise call it by many names) and *Vasudhaiva Kutumbakam* (The Cosmos is one family). On the basis of this they claim Hinduism as a universalistic and tolerant religious philosophy.

For Hedgewar, *Hindu Rashtra* is a cultural and emotional concept, eternally asserting itself. It makes clear that national sentiment is supreme, and all other sentiments, whether religious or sectional, should remain subservient to it. Hedgewar emphasizes that Hindu culture is the national culture, Hindu society is national society, Hindu art is national art, Hindu literature is national literature and Hindu history is national history. People belonging to various religions in this country are all Hindus by culture and are nationals of this *Hindu Rashtra*.

REFERENCES

1. Nisbet, Robert A., *Community and Power* (Oxford: Oxford University Press), 1968, p. 7.
2. Gupta, Dipanker, *The Context of Identity: Sikh Identity in a Comparative Perspective* (Delhi: Oxford University Press), 1996, p. 141.
3. Kakar, Sudhir, *The Colours of Violence* (Delhi: Viking), 1995, p. 247.
4. Purani, A.B., *The Life of Sri Aurobindo: A Source Book Pondicherry,* Sri Aurobindo Ashram, 1964, p. 81.
5. Raju, P.T., *Idealistic Thought in India* (Cambridge: Harvard University Press), 1953, pp. 440-441.
6. Kothari, Rajni, *Politics in India* (Boston: Little Brown & Co.), 1970, p. 34.
7. Paramananda, Bhai, *Hindu Sangathan* (Lahore: The Hindu Yuvak Sabha), 1936, p. 9.
8. Savarkar, V.D., *Hindutva* (Poona: S.P. Gokhale), 1949, p. 67.
9. Golwalkar, M.S., *Bunch of Thoughts* (Bangalore: Jayarana Prakashana), 1980, p. 73.
10. *Ibid.*
11. Chritophe Jaffrelot, "Hindu Nationalism: Strategic Syncretism in Ideology Building", *Economic and Political Weekly*, XXVIII (12-13) 20-7 March 1993, p. 517.
12. Wolport Stanley, *Tilak and Gokhale: Revolution and Reform in the Making of Modern India* (Bombay: University of California Press), 1962, pp. 64-65.
13. Ibid., pp. 178-79.
14. Savarkar, V.D., *Hindutva, op. cit.*, p. 67.
15. *Ibid.*
16. Keer, Dhananjay, *Veer Savarkar* (Bombay: Popular Prakashan), 1966, p. 265.
17. Golwalkar, M.S., *We or Our Nationhood Defined* (Nagpur: Bharat Publications), 1939.
18. Golwalkar, M.S., *Bunch of Thoughts, op. cit.*

4

MAHATMA GANDHI AND CONGRESS

Hedgewar now delved deep into the various movements going on at the time for the country's emancipation. He kept continuous contacts with Lokamanya Tilak, Dr. Moonie, M.S. Aney and other national stalwarts. He worked in responsible positions in the Congress and Hindu Mahasabha.

He was appointed a joint secretary of the Provincial Congress in 1922. He noticed pronounced indiscipline among the workers at every level. To counter it he tried to organise a disciplined volunteer corps as part of the Congress. To start with, he selected four voulunteers from each taluk. But the experiment did not yield the expected results. Volunteers in those days were expected to merely act as 'charge-free hamals' and play second fiddle to leaders. Besides, those who had taken the Gandhian vow of non-violence were basically opposed to such an organization. Dr. N.S. Hardikar of Hubli whom Hedgewar has known from his student days had started the Hindustani Seva Dal in 1923 during the Kakinada Congress session. Dr. Hardikar was supported by Pandit Nehru. Referring to Dr. Hardikar, Pandit Nehru wrote in his Autobiography: "We were surprised to find later how much opposition there was to the Seva Dal among leading Congressmen. Some said that this was a dangerous departure, as it meant introducing a military element in the Congress, and the military arm might overpower the civil authority! Others seemed to think that the only discipline necessary was for the volunteers to obey orders issued from above, and for the rest it was hardly desirable for volunteers even to walk in step. At the back of the mind of some was the notion that the idea of having trained and drilled volunteers was somehow inconsistent with the Congress principle of non-violence."[1]

Hedgewar used to be the most active participant in all national meets and conferences. He presided over a Congress conference at Dehgaon village in April 1923. He had great reverence for Gandhiji, in spite of his sharp differences with him in regard to several of his policies. When Gandhiji was arrested and convicted for six years on 18th March 1922, the 18th of every month came to be observed as Gandhi day till his release. On that occasion in October, this is what Hedgewar observed: "This is a most sacred day for us, for it is an occasion for us to listen to and contemplate on the noble virtues of the saintly life of Mahatmaji. And those, especially, who call themselves as his followers, have a special responsibility of emulating those qualities in their day-to-day life. The greatest virtue of Gandhiji lies in his capacity to dedicate himself wholly and solely for the nation's cause. To preach one thing and practice the opposite, would be the very antithesis of what Gandhiji stands for. To shout 'Mahatma Gandhi ki jai' and raise one's hand in support of his programmes in the public meetings and then coolly walk back to one's house and get totally lost in social affairs would be nothing less than hypocrisy. Let us not chant the hymn of peace just to cover up our weakness. Let us first become physically stronger than our adversary and then speak the language of peace. It is the only he who is prepared to throw himself and his family into the sacrificial fire of struggle that can rightly claim to be a follower of Gandhiji."[2]

Such was the beautiful harmony he was building up between diametrically opposite viewpoints. He never felt any contradiction in his participating in the non-violent *Satyagraha* launched by Gandhiji and following all its rules, even though he himself was a fiery revolutionary who believed that it was armed rebellion alone which would ultimately free the country from the shackles of foreign slavery.

The Non-cooperation Movement gradually died down; Gandhiji had been jailed. Many who had jumped into it in a momentary flush of enthusiasm were now disillusioned; they now sulked. On 5th February 1922, people in Chauri Chaura in Uttar Pradesh made a violent assault on the police station. They murdered 21 policemen and one officer, and set fire to the police station. Gandhiji was shocked by the incident. He called off the Non-cooperation Movement on 12th February.[3]

In all, over twenty thousand people had participated in the movement. What had prompted most of them was the catchy slogan "Independence within one year." People therefore became dispirited when the movement came to a grinding halt. Gloom and disappointment stared them in the face. Pandit Nehru, after he came out of the prison in 1923, described the situation thus: "We felt exhilarated, but this was passing sensation, for the state of Congress politics was discouraging enough. In the place of ideals there were intrigues, and various cliques were trying to capture the Congress machinery by the usual methods which have made politics a hateful word to those who are at all sensitive."[4] In such an environment, students who had forsaken their studies naturally went back to their books. Lawyers who had boycotted the courts recommenced their practice.

Even though the Non-cooperation Movement was directed against the government, it is clear that Gandhiji worked constantly with one eye on Hindu-Muslim Unity. He therefore set about trying to woo the Muslims by fully supporting the Khilafat Movement. But Hedgewar sensed danger in that move. In fact, he did not relish the newfangled slogan of 'Hindu-Muslim Unity.' He once met Gandhiji to explain his viewpoint. Without mincing words, Hedgewar questioned Gandhiji: "There are in India people of various religions like Hinduism, Islam, Christianity, Zoroastrianism, Judaism, etc. Why then do you harp upon only 'Hindu-Muslim Unity' instead of promoting the unity of all the various sects?" Gandhiji replied, "This will secure the friendly attitude of Muslims toward our country. As you are witnessing, they can be persuaded to side wholeheartedly with us in the nationalist struggle."

Hedgewar said: "Long before this slogan of Hindu-Muslim Unity came into vogue, many leading Muslims had identified themselves with our nation and worked shoulder to shoulder with Hindus under the leadership of Lokmanya Tilak, e.g., Barrister Jinnah, Dr. Ansari, Hakim Ajmal Khan and others. But I am afraid that this new slogan, instead of helping unity, might further aggravate the feeling of separateness among the Muslims."

"I have no such fear," said Gandhiji and closed the issue.[5] But, unfortunately, events of later years amply testifies to Hedgewar's fears.

When Mahatma Gandhi launched the 'Salt Satyagraha' in April 1930, Hedgewar was naturally pleased. Many workers of the Sangh sought permission to participate in the movement. But Hedgewar had undertaken the enduring and basic task of national rejuvenation through the Sangh. That long-term work had to go unhindered and grow stronger whatever the external conditions. Hedgewar wrote to all the Shakhas: "The Sangh as such has so far not resolved to participate. But the swayamsevaks in their personal capacity are free to take part after obtaining permission from their Sanghachalaks."[6]

A memorable event in the year 1934 was the visit of Mahatma Gandhi to the Wardha District Camp of the Sangh and the meeting between Gandhiji and Hedgewar. Gandhiji was then camping at Sevagram Ashram near Wardha. The Sangh camp was right opposite that house. Near about 1,500 Swayamsevaks of Wardha district were camping. Gandhiji had seen the preparations for setting up the camp. He also observed the camp activities from his residence and felt like seeing the camp from close quarters. He spoke to Mahadev Desai who in turn wrote to Appaji Joshi. Appaji immediately went and met Gandhij and invited him:

"Please do come whenever it is convenient for you. You will be most welcome."

That was the day of silence for Gandhiji. He wrote on a slip of paper "I will come at 6 tomorrow morning and shall spend an hour and a half there."

The next day, i.e. on 25th December 1934, Gandhiji arrived at the Sangh camp punctually at 6 a.m. All the Swayamsevaks offered their *pranam* to him. Gandhiji was accompanied by Mahadev Desai, Mirabehn and a few others. After seeing the camp in its impressive get up, Gandhiji patted Appaji on his back and said: "I am really delighted. I have not witnessed such a sight anywhere in the country before."[7] He then inspected the kitchen. Food was prepared and served to over 1,500 people without the least fuss or confusion ; the cost to each Swayamsevak came to just one rupee and a certain quantity of grain, and all the expenses of the camp were met by this; the Swayamsevaks themselves made up the shortfalls, if any. All this information astonished Gandhiji.

He later visited the sick ward and the tents of the Swayamsevaks. He saw people of all castes and classes harmoniously engaged in work without any reservations. In order to confirm his observation, he interrogated the Swayamsevaks themselves. Their response was uniform : "There are no differences like Brahmin, Maratha, Aspris, hyas, etc., in the Sangh. We are in fact not even aware of what castes many of our Swayamsevak brethren belong to; nor are we interested in knowing it. It is enough for us that we are all Hindus." Gandhiji asked Appaji, "It appears almost impossible to ward off the evil of untouchability from our society. How has this phenomenon been made possible in the Sangh?" Appaji replied, "Feelings of high and low, touchability and untouchability can be abolished only by emphasizing the inherent oneness of all Hindus. Then only will the spirit of fraternity be reflected in their sincere behaviour, and not merely in words. The credit for this achievement goes to Dr. Keshavrao Hedgewar."

By then it was time for the general assembly. The bugle was sounded. All the Swayamsevaks stood to attention. Flag was hoisted. In accordance with the Sangh custom, Gandhiji also joined Appaji in offering *pranam* to the Bhagawa Flag.

Gandhiji then visited the stores. Provisions and various requisites had been neatly arranged. There Dr. Hedgewar's portrait was decorated and prominently displayed. Mahatmaji looked at it intently and enquired, "Whose portrait is this?"

"That portrait is of Dr. Hedgewar," replied Appaji.

"Is that the same Dr. Keshavrao Hedgewar you mentioned when we were talking about untouchability ? How is he connected with the Sangh?"

"He is the Chief of the Sangh. We call him the Sarsangha-chalak. All the activities of the Sangh are carried on under his guidance. It is he who has started the Sangh."

"Will it be possible to meet Dr. Hedgewar ? If possible I want to hear about the Sangh from him directly."

"He is due to visit the camp tomorrow. If you desire, we shall bring him to you."[8]

Gandhiji then returned to his Ashram.

Next morning, Hedgewar arrived at the camp. That evening the valedictory function was held under the presidentship of Bhopatkar of Pune. Afterwards, at Gandhiji's invitation, Hedgewar went to meet him, accompanied by Appaji Joshi and Bhopatkar. The discussion between Hedgewar and Gandhiji extended for over an hour. The general trend of the discussion was as follows:

After preliminary exchange of courtesies, Gandhiji turned to a point uppermost in his mind:

"Doctorji, Your organization is admirable. I am aware of the fact that you were for many years a Congress worker. That being so, why did you not build such a volunteer cadre under the aegis of a popular organization like the Congress itself ? Why did you float a separate organization?"

"It is true that I worked in the Congress. I was also the Secretary of the Swayamsevak Dal at the time of the 1920 session of the Congress, when my friend Dr. Paranjpe was the Dal President. Subsequently the two of us tried to build such a volunteer cadre inside the Congress. But our efforts were not successful. Hence this independent venture", replied Hedgewar.

"Why did your attempt fail ? Was it for want of financial assistance?"

"No, no! There was no dearth of funds. Money can be a great help, no, doubt. But money alone cannot accomplish everything. The problem that faced us was not one of money but of attitude."

"Is it your opinion that noble hearted people were not there in the Congress, or that they are not there now?"

"That isn't what I meant. There are many well meaning people in the Congress. What is an issue is certain basic attitude. The Congress has been formed primarily with a view to achieving a political end. Its programmes have also been drawn up accordingly, and it needs volunteers to arrange for these programmes. The Congress leaders are therefore used to looking upon volunteers as unpaid servants who arrange chairs and benches during meetings and conferences. The Congress does not seem to believe that the problems of the nation can effectively solved only when there is a large and disciplined body of dedicated Swayamsevaks who are eager to serve the country of their own accord and without waiting for inspiration from elsewhere."

"What exactly is your conception of a Swayamsevak?"

"A Swayamsevak is one who would lovingly lay down his life for the all-round upliftment of the nation. To create and mould such Swayamsevaks is the aim of the Sangh. There is no distinction between a 'Swayamsevak' and a 'leader' in Sangh. All of us are Swayamsevaks and are therefore equal. We love and respect everybody equally. We give no room for any differences in status. This is in fact the secret of the remarkable growth of the Sangh in such a short period without any outside help, money or publicity."

"I am indeed very glad. The country will certainly be benefitted by the success of your efforts. I have heard of the vast following the Sangh has acquired in the Wardha district. ... How do you meet the expenses of such a huge organization?"

"The Swayamsevaks themselves bear the burden, each offering his mite as *Gurudakshina*."

"It looks as if your entire time is consumed by this work. How do you carry on your medical profession?"

"I have not taken to medicine as a profession."

"How then are you supporting your family?"

"I am not married."

Gandhiji was evidently taken by surprise. In the same surprised tone he said, "I see—you are not married ! Very good. That explains the remarkable degree of success you have achieved in such a short duration !"

At the end, Doctorji said, "I have probably taken too much of your time. I have no doubt that, with your blessings, our efforts will succeed. Permit us to take your leave now."

Gandhiji came up to the door to bid farewell, and said, "Doctorji, with your character and sincerity, there is no doubt you will succeed."[9]

The conversation shows and otherwise also, Hedgewar had great respect for the Mahatma. But the three incidents, viz., the Moplah Riots, the Khilafat Movement and the denunciation of the

Vande Mataram had disturbed him most. However, it was the third incident, i.e., the betrayal shown by the Congress towards the martyrs and the heroes of the nation by promising to drop certain stanzas from the National Anthem, which had set his head reeling. To kill the national spirit at the behest of the Maulana just to appease Muslim fundamentalists was nothing but a cruel joke with the nation, he thought. If the Congress could tolerate the insult to the National Anthem at the very hands of its President, Maulana Mohammad Ali, at the Kakinda Session, could it not easily compromise with the national ethos without remorse? If the Congress could think of killing its own heroes, could it not go to any length in future in the game of power politics, even to the extent of selling the nation to powerbrokers without any pangs of guilt? Surely, neither national heroes nor national symbols nor national ethos were safe in the hands of the Congress, Hedgewar concluded.[10]

He recalled how shabbily Shraddhananda was treated over the question of the Shuddhi Movement. "On the 13th February 1923," wrote the Swamiji, "I was called to lead the movement for the reclamation of Malkana Rajputs by their several brotherhood.... I found to my astonishment that while Mohammedan leaders doing *tabligh* (conversion) work openly, were allowed to guide the policy of the Congress and work as its accredited representatives, those engaged in the work of rescuing the Hindu Samaj from disintegrating were tabooed, and kept out of the Congress executive."[11]

Hedgewar knew why and how the Congress was born in 1885 to watch the interests of the British more than those of the Indians. No doubt, with the passage of time and the advent of leaders like Aurobindo, Tagore, Tilak and Gandhi, it had gone somewhat national in its outlook but at the bone marrow level it was supposed to carry the inherent seeds (*sanskars*) of disloyalty with which it was born; and which could play a mischievous role any time in future. It was this dubious role of the Congress in torture which disturbed him most because the history of its birth and the role it had played in the early years did strengthen his devotion. Surely, the future of India was not secure in the hands of the Congress in the long run because of the dubious germs it carried from its very birth.

The history of the birth of the Congress and the role played in the first two decades may be summed up as follows:

"The year 1885 is a very important year in the history of modern India because in this year was born (better say, given birth), on December 28, 1885, an organisation by the name of Indian National Congress. It was created as a "safety valve" to release the pressure of discontentment of the Indian masses against the British Raj. Allan Octavion Hume, an English I.C.S. officer, who had gone through the nerve wrecking days of 1857 with his ears pinned to the ground clearly heard the rumbling underneath. Also the reports coming to him from different parts of the country left him in no doubt that something very dangerous and highly explosive was simmering at the bottom, and it might burst forth if urgent remedial steps were not taken. The signs were the sudden outbreak of violence, sporadic crimes, murder of high English officials, bank robberies, looting of government treasuries, etc. Hume consulted Governor General Ripon and his successor Lord Dufferin on this issue and it was mutually decided that an organisation should be launched urgently to avoid the imminent danger, and so the Indian National Congress was born with the active co-operation of Indian loyalist bureaucrats in 1885, with Hume as one of its founding fathers."

Even Hume candidly expressed the underlying motive in his own words: "A 'safety valve' for the escape of great and growing forces, generated by our own action, was urgently needed and no more efficacious safety valve than our Congress Movement could possibly be devised."[12]

The official history of the Indian National Congress by Dr. Pattabbi Sitaramayya has also vouched for this background in the founding of the Congress. He wrote: "Mr. Hume had unimpeachable evidence that the political discontent was going underground. He came into possession of seven volumes containing reports of the seething revolt incubating in various districts, based upon the communications of the disciples of various gurus to their religious heads .... Hume thereupon resolved to open a 'safety valve' for this unrest and the Congress was such an outlet".[13]

This "safety valve" worked wonderfully well for the first two decades after its birth. President after President of the Congress sang the glories of the British Raj. The first President, W.C. Banerjee, opened the first session of the Congress in 1855 with prayers for the long life and the health of the Queen of England and spoke very highly of the noble character of the English race in general and of the British Government in India in particular. He talked about the varied and many benefits that were flowing and would continue to flow to a "poor" country like India under the benign rule of the British.

The second President of the Congress, Dadabhai Naoroji, exclaimed with warmth and pride that "they were laying the second foundation stone for the British Empire." Surendranath Banerji, presiding over the 1895 session, prayed that the British Empire might last forever, and again in 1902, he stated that "the highest aspiration of Indians was that they be admitted into the great confederacy of self governing states of which England was the august mother."[14]

Bengal changed the course of the future history of the Congress. It was the upheaval of 1905 that changed its contours and put on its pro-British body the garment of pro-Indian nationalism. But beneath the garment were dubious elements which could go off the national hook any time and play mischief with the destiny of the nation. It was this fear which disturbed Hedgewar the most.

Hedgewar was also doubtful about the ideology of the Congress. He saw that the Congress party was not a party of disciplined soldiers wedded to one ideology but a conglomeration of souls of different hues coming together on a common platform which was open to all. Anybody and everybody could come and go at will without scruples, could clash and shout at one another when their own interests local, regional, personal or communal-clashed. He thought of infusing discipline in the party once or twice when he was an active member of the Congress. He had organised a disciplined volunteer corps in 1922-23 as a part of the Congress, but his experiment was rebuffed by the High Command on two grounds. First, if such an experiment succeeded, it would end up in militant nationalism which went against the concept of non-violence, and secondly, such an organised body of young disciplined soldiers could refuse to toe the line of the High Command or play second fiddle to the bosses.

He also saw that the Congress was riddled with one ideological problem from the very day of its inception. And for that very reason it had to adopt the "Policy of Appeasement" towards Muslims. The problem was that the Congress could not be considered a national body by the British Government until and unless it came to represent all the religious communities. In other words, for the Congress to be national in character in the eyes of the British Government, all the communities, especially Muslims, had to take part in its activities. Thus, the problem with the Congress was to seek cooperation from all communities, especially Muslims, whether they came forward or not; and then and then only the British Government would look into the demands of the Congress. To seek that cooperation from the Muslims—in order to be considered national—the Congress could go to any extent even to the extent of being ridiculous. For example, it paid for the fare and other expenses of Muslim delegates travelling to the Congress sessions. Surendranath Banerji noted: "Our critics regarded the National Congress as a Hindu Congress, and the opposition papers described it as such. We were straining every nerve to secure the cooperation of our Mohammedan fellow-countrymen.... We sometimes paid the fares of Mohammedan delegates and offered them other facilities."[15]

Swami Shraddhananda, a great social and religious figure and an eminent leader of the Congress in those days, had given some interesting details of the 1899 Congress Session at Lucknow:

"Tickets were issued to every Muslim '*waiz*' delegate free of charge. Messing fee of Rs. 10 per head, too, was not charged from the Muslim delegates while they were served with all the delicacies of 'Dastar Khan'...And these '*waiz*' delegates stopped in the Pandal only a few minutes in the beginning and were to be found enjoying creature comforts under the refreshment *shamianah* outside the pandal for the rest of the sitting..."

"The majority of Muslim delegates had donned gold, silver or silk embroidered *chogas* (flowing robes) over their ordinary coarse suits of wearing apparel. It was rumoured that these *chogas* had been lent by Hindu moneyed men for Congress *tamasha*. And of these the majority were admitted free to delegates' seats, board and lodging. Sir Syed Ahmed's anti-Congress League had tried in a public meeting to dissuade Muslims from joining the Congress as delegates. As a counter move, the Congress people lighted the whole Congress camp some four nights before the session began and advertised that ingress that night would be free. The result was that all the *chandu khanas* (opium dens) of Lucknow were emptied and a huge audience of some thirty thousand Hindu and Muslim delegates were elected or selected...."[16]

At the level of constitution also the Congress tried to appease Muslims. In its Allahabad session of 1888 under the Presidentship of Tyabji, it adopted a resolution which declared that the Congress shall not discuss any fresh subject or pass any fresh resolution which the Hindu or Mohammedan delegates opposed unanimously or near unanimously.

This amounted to granting the power of veto to Muslims, however small their numbers might be in the Congress, to torpedo any policy or programme of the party. Having accepted this slippery position, the Congress started its downhill journey with increased speed as the days passed.

This was how the Congress, Hedgewar thought, began bending over backwards to please the Muslims and let them go ahead with their divisive tendencies. Instead of giving them lessons in emotional integration, in being one nation and sharing common national aspirations and common joys and sorrows, it encouraged them to be separate in all walks of national life, especially in politics.

The fate of the nation, according to Hedgewar, was sealed when the Moderates in the Congress under the leadership of Gokhale gave their assent to the Minto Morley Reforms 1909 under which "special weightage and special electorates" were guaranteed to Muslims. This was the beginning of the 'Two-Nation Theory", the seeds of which the Moderates had sown just to oust the Extremists from the show.[17]

When the Congress-League Pact 1916 or the Lucknow Pact was on the anvil, the Congress gave a greater share to Muslims in Provincial Assemblies than what was agreed upon in the Minto Morley Reforms 1909. The background was that it (Congress) wanted to snatch the initiative from the hands of the British Government by going one step ahead in pleasing the Muslims. So it thought of entering into a Pact directly with the League without enlisting the support of a third party. That was how the Lucknow Pact of 1916 was born. However, the most dangerous proviso that was added to the Pact was:

"That no bill, any clause thereof, nor any resolution, introduced by a non-official member affecting one or the other community in the Legislative Council concerned, shall be proceeded with if three-fourths of the members of the community in the particular council, Imperial or Provincial, oppose the Bill, any clause thereof, or the resolution."[18]

In other words, the Congress by agreeing to the above clause had virtually given veto power to Muslims to negate or drop any clause, resolution or programme which did not favour them or suit their taste. Also through the Lucknow Pact, the Congress had given sanction in black and white to two major and pernicious doctrines: one, the right of Muslims for separate communal electorates and communal representation and two, the claim of the Muslim League to speak for the entire Muslim community in India.

The Government lost no time in putting the official seal on the Lucknow Pact 1916 by indirectly incorporating all the important features of the Pact in the matter of Muslim representation in the Constitutional Reforms 1919—thus sealing the fate of One-Nation Theory—India as One Nation on the Indian soil permanently.

Then came the days of the Khilafat. It was here that all the reasonable limits that could contain the 'Appeasement Policy' were thrown to the wind indiscreetly. It was the height of nonsense on the part of the Congress to enter into the Khilafat Movement just for the sake of Khalifa of Turkey and push the country into rivers of blood that flowed from the springs of communal violence. The poor Indians believed in what Gandhiji had said: "The Khilafat question gave the two communities the opportunity of a lifetime to unite. If the Hindus wish to cultivate eternal friendship with Musalmans, they must perish with them in the attempt to vindicate the honour of Islam."[19] And the Hindus actually perished with the Muslims. But to what effect? When the Khilafat failed, Gandhiji became a *Kafir* overnight, the Hindus, the worst enemy of Islam, and there broke out riots everywhere, the Malabar riots being the worst.

But the worst thing that the Khilafat Movement did was that it set aflame the spirit of extra territorial nationalism in the minds of the Muslims of India. They began to think of themselves as a part of one big Muslim world—Pax-Islamia—instead of considering themselves a part of India. In other words, they began to put their religious loyalty over and above their national loyalty.

From that day onwards began the process of disintegration, Hedgewar thought. And what would be the end of this slipping trend, down and down if it was not arrested and the situation corrected urgently?

In his dairy he wrote, "the upsurge witnessed during the days of Non-cooperation Movement had died down. The various evils accompanying the movement are now having a heyday. Mutual distrust and ill-will, personal and caste rivalries, have all raised their ugly heads. No institution or organisation seems to be free from these internal squabbles and utter lack of discipline. The snake of Muslim fanaticism, having been fed on the milk of the Khilafat, is now baring its poisonous fangs and spreading the venom of violence and riots all over the country."[20]

Gandhiji enjoyed a unique position in national politics. Hedgewar had a soft spot for him, especially for the ideal Gandhi who, after a long ordeal of 21 years in South Africa, came to India in 1915 with pious hope of changing the destiny of his mother-country. He also appreciated his austere life, truthful nature, and above all his full and final commitment to the concept of *Purna Swaraj*, the concept which he had expounded in a small booklet entitled *Hind Swaraj*. But where he differed with Gandhiji or failed to see eye to eye with him was on (1) his vision of Hindu-Muslim Unity; (2) his doctrine that Swaraj could not be won without the cooperation of Muslims; and (3) his creed of 'Non-Violence' that must be observed at all costs and under all circumstances.

So far as Gandhiji's vision of Hindu-Muslim Unity was concerned, Hedgewar felt it was going to end up in a sad fiasco because of the Appeasement Policy the Congress was playing. In

South Africa Gandhiji had succeeded in welding the two communities because of the situational demand that Hindus and Muslims should participate in the struggle against their common enemy, the 'Black Law', as Indians and not as members of two different communities. Gandhiji carried the same vision from the foreign land and planted it on Indian soil where the context was different.

In India, Hindus and Muslims lived and acted as members of two different communities and not as Indians. And many forces were at work to widen that rift further. For example, there were the British Government, the Muslim League, the Communist Party of India (CPI), and many vested interests within the Congress itself. Even Maulana Mohammed Ali, the right hand of Gandhiji during the Khilafat days, did not share his master's vision of Hindu-Muslim Unity. He was there with Gandhiji not for Hindu-Muslim Unity but for the restoration of the Caliphate, the institution which the Turks themselves had pulled down. When the Khilafat ended in smoke, the Hindus overnight became *kafirs* and Gandhiji a condemned politician, worse than a debauch. "Yes, according to my religion and creed," said Maulana Mohammed Ali in a public meeting in Lucknow in 1924, "I do hold an adulterous and a fallen Musalman to be better than Mr. Gandhi."[21]

So Gandhiji's vision of Hindu-Muslim Unity was nothing but a grand mirage in the desert of communal politics. Hedgewar thought it could not be achieved, especially in the way the Congress was pursuing. By showering undue concession without commitment, Hedgewar asked himself, could the heart of the receiver be ever changed? And if the exercise continued for long, he wondered, would it not sharpen the biting teeth of the latter, endangering not only the very life of the giver (Congress) but also of the nation as a whole. And if that situation arose, it would be a historic tragedy of the worst kind, he imagined.

So far as Gandhiji's doctrine, 'India could not win freedom without the cooperation of Muslims', was concerned, Hedgewar believed it was unwanted, unscientific and totally pernicious to the collective health of the nation. After coming in touch with the actual politics of the day in India, Gandhiji understood the Hindus were by and large with the Congress against the British, and the Muslims more or less with the Muslim League in favour of the British. From that scenario, Gandhiji had, perhaps, concluded that it was not possible to win freedom without the actual participation of Muslims in the freedom struggle. For that very reason, Hedgewar thought Gandhiji must have started feeling positively about appeasing Muslim interests. But the irony of the situation was that as he tried more and more to secure Muslim participation, and as their participation slipped more and more from his hands, he drifted more and more away from the commitment with which he had landed in India. The result was that under the influence of the said pernicious doctrine the Hindu started losing faith in his own vitality and strength to fight single handed for freedom, and the Muslim, on the contrary, started getting more and more aggressive towards the meek Hindus. And if that scenario continued to last long, Hedgewar feared, it could result in undermining the self-confidence of the Hindus on the one hand, and strengthening of the aggressive attitude of the Muslims on the other. The Hindus could lose their self-resolution in fighting and winning freedom on their own strength, and the Muslims gain the benefit of demanding their pound of flesh from the body of Bharat Mata as a *quid pro quo* for participating in the struggle for freedom, as they had started feeling themselves a separate nation under the influence of the Khilafat. The whole game might end up one day in fracturing the very concept of the 'One-Nation One-Country.'[22]

Gandhiji believed in changing the heart of the foe with the power of LOVE—a high flown platonic axiom! "Heal hate by love" —this is what Gandhiji used to say to his followers. But 'love' that heals 'hate' is not an ordinary kind of love, the kind which is heard and found in common parlance in the world of modern-day lovers. It is love of a very high degree, say, of divine nature, the kind which comes to possess one when he is bathed in the sun-shine of oneness with the Supreme Soul of the entire universe. It is love, for example, that flows from the heart of a saint like a fountain for the entire mankind, nay, for the entire life—man, bird and beast—and whose heart bleeds when he sees even when the nature is cruelly treated and weeping. Or, the kind of love that comes to possess a philanthropist who is consumed with desire to heal the miseries of the race even at the cost of his life. But the mass of humankind does not and cannot rise to the height of a saint or an honest philanthropist. To ask the masses of humankind to act as saints or rise to the height of divine love and practice it in relation to their adversaries or oppressors is to ignore the human nature. Practically it amounts to giving a free hand to injustice and violence by expecting that the hand of the deliverer would be paralyzed when raised to strike under the force of ordinary love of common parlance.

Gandhiji also wanted his followers to accept on faith his creed of 'non-violence' a creed which they must follow at all costs and for all times and in all situations. First, for Gandhiji to demand such exacting endurance from his followers—that they all must suffer to the last man and the last woman without raising a finger at their tyrant—was asking for the moon (proverbially speaking) because it was altogether out of proportion with human nature. Secondly, it was wrong for him to raise 'Non-violence' from a pragmatic and moral credo to a fetish and a fundamentalist credo and that too for political ends. Thirdly, it amounted to philosophical corruption when he argued that "military spirit had never been a part of Hindu tradition."

No doubt, 'non-violence' had been (and is) an accepted part of the Hindu way of life, but it had never been elevated to a presiding deity before which everyone must bow under all circumstances. Hindu *dharma* had never accepted non-violence as creed for everyone to observe at all costs. It had never allowed it to be changed into a hardened dogma which everyone must follow even at the cost of honour of his mother, sister or his country. It is to be noted here that all Hindu gods and goddesses are possessed with special weapons and warheads. Lord Krishna had to force violence on the battlefield of Kurukshetra when the evil mind failed to see the light of reason. The leaders of Hindu Renaissance like Mahrashi Dayananda, Swami Vivekananda and Sri Aurobindo had advocated violence when the forces of evil could not be contained through peaceful means. In the First World War, Tilak, Annie Besant and other ardent nationalists had demanded political freedom on a condition precedent to India's unreserved participation in the war. And now to accept Gandhiji's creed of 'non-violence' as faith (as he had wished while terminating the Civil Disobedience and Non-Cooperation Campaign 1922 on the issue of the Chauri Chaura case) meant going against the ethos of the nation. If Arjuna, the wielder of the Gandiva, was asked to fight without his Gandiva, Hedgewar wondered, how could he lead the forces of *dharma* to victory?

'Non-violence' is a weapon of the strong; it is not a weapon of the weak. Honestly speaking, it is to be observed by those who have the strength to punish the wrong doer. For the weak to observe it amounts to exposing, their suicidal helplessness to the tormentor who draws sadistic pleasure from the pain. For Gandhiji to ask people to observe 'non-violence' of his creed under the political firmament, which ran full to the brim with the stuff called fraud, chicanery and viciousness, amounted to throwing his sheep before the wolves, i.e., encouraging the British Government on the one hand to unleash their repressive measures unchecked and the Muslim League on the other to let loose its goonda-hounds on the meek Hindu, asking them to go on a biting spree without reservation and free of danger, and then allowing them all to get off unpunished and unharassed in the end after perpetrating the heinous crimes. What was needed was that Gandhiji should have strengthened the fibre of the people first with moral as well as physical grit before asking them to face the perpetrator with the weapon of 'non-violence', said Hedgewar.[23]

REFERENCES

1. Nehru Jawaharlal, *An Autobiography* (New Delhi: Oxford University Press), 1982, p. 121.
2. Seshadri, H.V., *Dr. Hedgewar: The Epoch Maker, op. cit.*, p. 16.
3. Sharma, Jai Narain, *Power, Politics and Corruption* (New Delhi: Deep and Deep), 2005, p. 130.
4. Nehru Jawaharlal, *An Autobiography, op. cit.*, p. 98.
5. Seshadri, H. V., *Dr. Hedgewar: The Epoch Maker, op. cit.*, p. 59.
6. *Ibid.*, p. 110.
7. *Ibid.*, p. 140.
8. *Ibid.*, p. 141.
9. *Ibid.*, p. 144.

10. Hingle, G.S., *Hindutva Reawakened, op. cit.*, p. 90.
11. Malkani, K.R., *The RSS Story* (New Delhi: Allied), 1984, p. 17.
12. Wedderborn, William, *Life History of A.O. Hume* (New Delhi: Pegasus), 1974, p. 77.
13. Sitaramayya, Pattabhi, *The History of the Indian National Congress*, Vol. I (Delhi: S. Chand & Co.), 1969, p. 11.
14. *Ibid.*, p. 19.
15. Seshadri, H.V., *The Tragic Story of Pakistan, op. cit.*, p. 51.
16. Shradhananda, Swami, *Inside Congress* (Bombay: Phoenix), 1946, pp. 25-33.
17. Hingle, G.S., *Hindutva Reawakened, op. cit.*, p. 95.
18. Mehta, Ashok and Achyut Patwardhan, *The Communal Triangle in India* (New Delhi: Oxford), 1942, p. 34.
19. Kulkarni, V.B., *India and Pakistan* (New Delhi: Sterling), 1988, p. 219.
20. Seshadri, H.V., *Dr. Hedgewar: The Epoch Maker, op. cit.*, p. 75.
21. Majumdar, R.C., *History of Freedom Movement*, Vol. 3 (Calcutta: Firma K.L. Mukhapadhayaya), 1966, p. 336.
22. Hingle, G.S., *Hindutva Reawakened, op. cit.*, p. 99.
23. *Ibid.*, p. 101.

5

THE TECHNIQUES THAT SUCCEED

An ideal requires a method, a process for its realisation in actuality. Today all around us we witness various types of techniques being adopted by several organisations. Even at the time of the founding of RSS these different methods of work were in vogue. And it was not that Hedgewar, the founder of the Sangh, was unaware of those systems when he evolved this particular technique. On the contrary, for quite a long time before he started the Sangh, he was in the thick of the activities of those organisations and was well acquainted with their techniques.

Why then did he start a separate organisation with a separate technique? Is it because he wanted credit and applause for himself as the founder and leader of an independent organisation? Nothing could be farther from truth. Hedgewar had received enough applause on other platforms. With his spirit of boundless sacrifice, tremendous energy of action and flash of genius it was mere child's play for him, had he so desired, to get the highest political prizes of the land. But he gave up all that and started this silent and unassuming type of organisation.

In Search of the Right Path

What then is the reason that prompted him to plough this fresh furrow in the national field? Being of a fiery patriotic temperament right from his boyhood, his first fascination was naturally for the revolutionary movements of the day. He had, as a matter of fact, chosen the medical course at Calcutta in spite of untold personal hardships only with a view to diving deep into that movement, as Calcutta was then the seething volcano of revolutionary activities. He lived in the dangerous under currents of that volcanic lava, but kept his discerning eye ever wide open. Though his heart throbbed in unison with the flaming hearts of those revolutionary comrades, he found their method wanting as an effective instrument for total national regeneration. Their secret and lightning movements, their daring exploits and glorious martyrdoms evoked the highest admiration in his fiery bosom but his calm brain refused to be blinded by the flash of such revolutionary sparks.

He knew that a handful of secret workers, deprived of direct contact with the people, could play but a very limited role in rousing and organising a whole nation. Further, he had observed that most of the revolutionary plots and secrets had met with disaster by the cropping up of informers and traitors every now and then from even amongst their top cadres, thus undoing at one stroke the glorious heroism and sacrifices of countless revolutionaries. Without a trained core of extremely strong-willed, disciplined and patriotic men, no revolution could be expected to succeed in blowing up an organised and mighty state machinery such as the British had built in our country.

After Hedgewar returned to Nagpur from Calcutta, he plunged into the freedom movement of the Congress. Under the leadership of Lokmanya Tilak, the Congress was fast becoming a movement of the masses. The age of arm chair politicians was gone. A new era of mass resistance to the British had set in. The spirit of swaraj was in the atmosphere. After Tilak, Mahatma Gandhi spread the fire far and wide and lit up the torch of resistance in every town and village. Gandhiji with his unimpeachable character, his spirit of utter selflessness and fearlessness, his simple yet effective techniques touching the heart of the common mass of people, roused the nation to new heights of struggle and sacrifice. Pandit Nehru with his burning idealism and dynamism was coming up as the inspiring symbol and spearhead of the proud and independent spirit of the rising generation. By the efforts of such stalwarts the country was caught in a wave of intense anti-British upsurge. Hedgewar too was in the forefront, bearing the brunt. He defied the foreign rulers and courted hard terms in prison. At the same time, he continued his close scrutiny of the movement and found to his disappointment some grave defects which he feared would be the undoing of all its cherished aspirations in the long run.

In both the movements he had observed that the main incentive was the anti-British spirit with no positive vision of national freedom. That was the case with not only the general mass of the people but with most of the leaders as well. And without a positive conviction of our national life, those movements, in course of time, were bound to drift into reactionary channels.

Secondly, they could give no answer to a question which had haunted Hedgewar since his very boyhood. How could a handful of foreigners, Muslim or British, subjugate the nation and rule over it for such a length of time? Ousting the British was all right, but what was the root malady which had resulted in that foreign domination? If, as it was clear, our own disunity was responsible for our slavery, then would these terrorist and agitational methods remedy that root malady? Could abiding oneness of the people be created by revolutionary shocks and mass upsurges? Could all the evil tendencies such as selfishness, lack of discipline and absence of national consciousness, which had been eating into the vitals of our people for the last so many centuries and had, resulted in foreign domination, be wiped out at one stroke?

And again, without the people being rooted in the positive and sustaining virtues of an organised national life, would it be advisable to work up mass fury on the basis of mere antagonism? In Shakespeare's Julius Caesar, after Caesar is assassinated by his own comrades, and Brutus and Cassius ride on the crest of popular upsurge, Caesar's friend Antony cleverly manipulates the mob fury and turns its tide against the assassinators themselves. As the wild mob turns in their hot pursuit, Antony comes out with the famous remark, "Mischief, thou art afoot. Take thou what course thou wilt." That is the disastrous drawback—'thou wilt' and not as 'we will'—of all such mass upsurges without a corresponding countrywide organisation based on positive national life values and capable of directing and controlling the movement.

For a Mental Revolution

After deep cogitation along these lines, Hedgewar concluded that a total revolution in the mental attitude of the people was the vital need of the hour. It was a mental revolution and not a physical revolution that was the panacea for all the nation's ills including the foreign domination. He was aware that the task of bringing about a total transformation in the attitudes and thought processes and behaviour of the whole people by taking individual after individual and moulding him for an organised national life, demanded a perseverant, silent and single-minded approach free from all public fanfare and propaganda.

National reorganisation means fostering those traits which build up national character and cohesion. It is directed towards awakening a passionate devotion to the motherland, a feeling of fraternity, a sense of sharing in national work, a deeply felt reverence for the nation's ideals, discipline, heroism, manliness and other noble virtues. This work of moulding minds and building character cannot be done by sermons or administering pledges. The spirit of devotion to the nation has to be a steady flame, burning day in and day out and year after year.

The Image for Worship

Together with its all-out emphasis on the practical aspect of the work, the Sangh has bestowed profound thought on evolving a proper framework of technique that will be effective in the achievement of its ideal. Without a suitable technique no ideal, however great, can be realised. Even in the various sects, each individual has a definite emblem in keeping with his particular sect. He dresses and adorns himself in a particular manner, recites a particular *mantra* and follows a particular code of discipline. A Shaiva, a Shakta or a Vaishnava, each has his own method of worship, his own ritual, his own codes and conventions regulating his life.

The Sangh have evolved a technique, an emblem, a *mantra* and a code of discipline in keeping with the ideal of an unified and disciplined national life. The great and inspiring emblem that they have chosen is the *Bhagwa Dhwaj* which brings before the eyes the living image of the ancient, sacred and integrated national life in all its pristine purity and entirely crossing all superficial barriers of province, sect, creed, caste, language and custom. Since times immemorial, it has been the symbol of the *dharma*, the culture, the traditions and ideals. It embodies the colour of the holy sacrificial fire that gives the message of self-dissolution in the fire of idealism and the glorious orange hue of the rising sun that dispels darkness and sheds light all around. It has been the one guiding star to all our endeavours, material as well as spiritual, the one unfailing witness to every page of history. It has witnessed the rise and fall of empires, the penance of yogis and the sacrifices of heroes and symbolised the dreams of countless millions of this land all through the ages. In short, it has been the highest, the noblest and the truest symbol of Indian nationhood.

Worship Ideal, Not Individual or Book

The Sangh has taken up this living symbol as its guiding light—the *guru*. When Hedgewar placed this flag before as the *guru*, as the ideal, quite a few of his co-workers raised their eyebrows. Having seen the ideal in flesh and blood in the form of Hedgewar himself, they queried, "Why not look upon Hedgewar himself as our ideal?" But Hedgewar in keeping with the spirit of the organisation, placed the immortal *Bhagwa* as the ideal. No individual, however great, can be the ideal for a nation. The individual is after all a fleeting entity in the eternity of national life. However great, he can at best reflect a fraction of the beauty and fragrance of the full bloom of the national life over the centuries. Moreover, it is futile to expect that all people will cherish the same devotion towards a particular person, however noble and venerable he may be. Some worship Sri Rama as their Chosen Deity whereas some others look upon Sri Krishna as their God and so on. Therefore the Sangh has kept a symbol which is at once universal and all absorbing in its appeal.

The tragedy of movements which revolved solely around individuals for their inspiration and ideals is there for all to see in the history of nations the world over. The curse of personality cult and the rise of dictators that have scarred the face of humanity have been due to the idolising of individuals to the neglect of ideals. The Indian culture has always commanded to look upon the individual as great and worthy of adoration only to the extent he expresses an ideal in his life. In the whole wide world it is the *dharma* alone that is not based on the historicity or authority of any single individual.

The other special feature of the heritage is that no book is taken as the single supreme authority for the *dharma and sanskriti*. All scriptural texts are only expositions of the several aspects and paths to the One Goal of human life. Hedgewar too has not accepted or prepared any book to serve as its sole authority. Once a prominent religious leader asked Golwalkar, "Which is the text you follow?" He replied, "If we confine ourselves to the word of a book, then we will be no better than Muslims and Christians whose religion stands on a book. And so our devotion is to the ideal and to nothing less as nothing else."[1]

It is in keeping with that sublime cultural tradition that the Sangh has kept before itself neither an individual nor a book as its authority but *Bhagwa Dhwaj*, the glowing symbol of all that is good and great in the national life, and through that, is striving for the inculcation of pure devotion to the nation as a whole.

Crucible of National Reorganisation

It is in the sanctifying presence of the *Bhagwa Dhwaj* that the day-to-day activities of the *Shakha* are carried on. All sections of the people gather there. Forgetting all superficial distinctions of language, province, caste, community, party or sect, they gather as children of a common motherland and play in her sacred dust. They pray to the motherland in deep veneration. They resolve to lay down their lives for her glory. As they play and sing, a feeling of oneness brings them together. As they perform exercises together and march together, their hearts begin to throb in unison.

More important than the programme is the atmosphere. An air of sweetness and sanctity pervades the atmosphere. In course of time, amidst the wide variety and diversity of the assembling persons, a wholesome unity emerges. The spirit of amity and harmony strikes root in their minds. And the inspiring dream of national unity submerging social, political, economic and other divisions becomes a living reality. Thus the Shakha is the crucible which awakens noble impulses of dedicated patriotic service in the people and binds them together with immortal fraternal bonds. It is the creative centre for sterling national character and lasting national cohesion.

It is clear from the above description that it is the all absorbing spirit of devotion to nation in its entirety and not to any individual or institution that is sought to be ingrained in the Sangh. Without that life spirit, mere attachment to and pride in the external set up will become one more point of national disruption.

In the past history and even in recent times, Indian have burnt the fingers having ignored this basic principle of national reorganisation. There had been many attempts in the past in various parts of the country at awakening and organising the people. The great sponsors of those movements had started them with a view to strengthening and unifying society as a whole. But as circumstances changed and the immediate cause for their coming into being was removed, the inner spirit gave place to mere attachment to the external form. As a result, one sees many such rigidified sects developing into mutually exclusive and even hostile entities in the present day atmosphere of selfishness.

For example, during the gruelling times of Muslim onslaught, there arose in Punjab a great saint Guru Nanak Dev who rekindled amongst the people the dying embers of faith in the ancient *dharma*. He was followed by a galaxy of nine gurus who lived and died as flaming examples of devotion and sacrifice in the cause of *dharma*. The tenth guru, Guru Govind Singh, felt that mere revival of *dharmic* devotion, without heroic action, would be of little avail against the brutal forces of *adharma*. He charged his followers with an indomitable martial spirit and forged them into a conquering army of heroic warriors. But what a misfortune that today the mission which inspired that glorious movement is given a go by and mere attachment to the external form, the institution, has so much hardened as to give rise to perverted notions of separatism laying the axe at the very life spirit with which it came into being![2]

Hedgewar has never entertained the idea of building an organisation as a distinct and separate unit within society. Right from its inception he has clearly marked out as its goal the moulding of the whole of society, and not merely any one part of it, into an organised entity. That is the reason for the Sangh workers' not parading themselves as a 'Sangh group' before the people, even when thousands of them work staking their all in times of national catastrophes like famines, floods, flow of refugees from Pakistan, etc. They are content to remain as ordinary members of society and thereby put an example of how even a common man should behave in an alert and organised social life. Such a well-knit, patriotic and self-sustaining national life alone can fortify the nation with overwhelming and everlasting strength.

The idea of building a powerful group within society—sometimes taking the form of a private army—is fraught with grave perils to a free and prosperous national growth. The world witnessed such bodies shooting up like meteors on the political horizon in Germany, in Italy, in Russia and China and establishing totalitarian regimes in their respective countries. It is in the nature of these totalitarian parties to seek to perpetuate their domination on society and to enslave people politically, economically, socially, culturally, and in all other respects. The nation's free expression is thereby choked. The individual is annihilated. And bereft of individual initiative and freedom, the society begins to degenerate.

The idea of domination through brute strength is absolutely alien to our culture and tradition. Numerous faiths and creeds have flourished here from ancient times. We have had a variegated pattern of political institutions. We have had republican governments and hereditary kingships. Under all conditions the people were free to follow their healthy persuasions in every walk of life. Everyone was encouraged to develop himself according to his individual genius, nature and inclination. In keeping with that spirit, the work of the Hindu missionaries for rousing and organising the society has always been through love and service, character and sacrifice and never through brute force or political power.

It is this type of elastic and self-expressive pattern of organisation that has helped this society to keep alive its spirit of coherence in spite of being subjected to unparalleled atrocities and aggressions. If the pattern had been rigid and imposed from above, the society would have today remained merely as a fossil, just as some of the huge animals became immobile and gradually perished under the dead weight of their rigid protective covers. Hedgewar has therefore rightly eschewed all such self-defeating alien types of organisation and stuck to the pure and healthy national system for rebuilding society.

The concept of total national reorganisation that the Sangh has been striving to bring into reality naturally implies the non-political nature of its work. After all, a political party can but represent a very small fraction of the people. Nor can national oneness be achieved through elections and political propaganda. Political techniques—and even political power for that matter—can hardly infuse the spirit of devotion, heroism, character, amity and sacrifice in the people. In fact, without having those grass roots of a well knit national life, the political parties degenerate into mutual hostility and ruin the national fabric.

That is the unfortunate predicament in which we find ourselves today. Devoid of the living faith in a single national entity and of the supreme consciousness of national interest we find that most of the political parties have become merely breeding centres for disruptive forces and fissiparous tendencies. We often find them joining hands with avowedly anti-national elements within the country and sometimes even with outside enemy powers in a bid to further their narrow party interests.

One may ask whether it is not desirable to have two parties working in a spirit of healthy competition so that each may act as a corrective to the other and thus keep the national health free from the poison of dictatorship of a single party or a single person. The democratic institution has undoubtedly that saving feature. But even that will come into play only when the people are firmly rooted in the consciousness of 'nation above party'. Even in those Western countries where the people are steeped in that particular tradition for the several centuries, parties are not entirely free from mutual rivalry and envy. However, they are able to keep the political pulls within limits and subservient to the higher call of national welfare.

For them such institutions do good and add vigour to their life. If a healthy and strong man puts on a shirt made of very fine mulmul, it will appear very nice for him. But if a person whose limbs are lean and thin, his chest gone in and who cannot even stand erect, wears that and tries to strut about, he only makes of himself a laughing stock in the eyes of others.

When Winston Churchill, then the leader of the Opposition in Britain, was touring in America, some persons put him questions about his attitude towards the Labour Government in his homeland. Churchill bluntly replied, "Abroad we are one, whatever our differences at home." Even during the two wars, in spite of the immense suffering and misery the people there had to undergo, no political party tried to make capital out of that situation for its own party ends. The repeated miserable failures of the Communists in that country to save even their deposits in the elections give us an indication of the Britishers' deep rooted sense of nationalism.[3]

The Lucid Contrast

It is out of a superficial view of the external set up of institutions in other countries without looking into their inner spirit that some persons ask us, "Countries like England have been progressing and triumphing over difficulties and ordeals even without any special effort such as yours. The people there are engaged in their normal routine life. The political parties carry on the affairs of the country quite successfully. Then, where is the need for a separate organisation such as yours in our country and for the day-to-day training that you are carrying on?"[4]

But there everyone is a born patriot. There is no need even to mention that a particular person is a patriot just as we do not add the prefix 'man' while referring to an individual. That he is a man is taken for granted. But in our country we often hear the word 'patriot' as a special word of tribute applied to certain persons. In England in every one of their institutions, whether a school or a college, a literary club or a youth league, a social meet or cultural body and even at home, the first lesson that every child learns is, "England! With all thy faults, I love thee still!"

And again see the inspiring tradition of patriotism which England has set up over centuries. During the whole of the past few centuries there were hardly any traitors. During the Second World War, there was but a single notable case of betrayal. The son of Lord Amery, who was then the Secretary of State for India, had worked for the Germans. After the war, he was tried and sentenced to death. His father would not even think of exerting the influence of his high office to get the punishment lessened. On the other hand, he declined to submit a mercy petition. On the day of his son's execution he even refused to see him, saying that it was a sin to see the face of one who was a blot on the glorious patriotic tradition of their family![5]

And here? A whole race of traitors, right from the times of Dahir and Prithviraj, down to the times British rule was born here. And that has continued unabated even to this day. Now-a-days treachery has become almost a passport for higher posts in our country.

In England, politics is a healthy sport for them. When the Labour Party was in power, it deputed as its ambassadors the Leader and the Deputy Leader of the Opposition Party to America and China, respectively. Can we ever dream of such a healthy convention on the part of the party in power in our country? Even the third rate and fourth rate persons of the ruling party are considered superior to the ablest in other parties. As a result, the entire administration is monopolised by the ruling party. They have not stopped even at that. The state machinery is freely exploited to serve their party interest. Agitations and movements by opposition parties are sought to be suppressed by brutal government measures.

With such germs of national disintegration eating into the vitals of our nation, it is useless to expect that mere copying of the political and other institutions of other countries will solve our problems and bring about an all-round national rejuvenation. Our malady is far deeper and requires a far more radical cure. It is to root out the basic malady that the Sangh has evolved the method of day-to-day training, the day-to-day inculcation of qualities such as the spirit of sacrifice, discipline and national devotion that go to build a resurgent and unified national life.

REFERENCES

1. Golwalkar, M.S., *Bunch of Thoughts* (Bangalore: Sahitya Sindhu Prakashan), 1966, p. 395.
2. *Ibid.*, p. 398.
3. *Ibid.*, p. 402.
4. *Ibid.*, p. 403.
5. *Ibid.*, p. 404.

6

FORMATION AND DEVELOPMENT OF RSS

The outbreak of communal rioting in 1923 caused Hedgewar to question the previously attempted methods used to rid India of colonial rule. The riots, in his view, were the signs of a deeper social problem —disunity among Hindus—that would not have to be addressed if India were to be independent. He observed the problem at close quarters since Nagpur was the one major of the centres of rioting.

During this period of escalating Hindu-Muslim animosity in Nagpur, Hedgewar began to develop the intellectual foundations of the RSS. A major influence on his thinking was a handwritten manuscript of Vinayak Damodar Savarkar's Hindutva, which advanced the thesis that the Hindus were a nation.[1] The Central propositions of Savarkar's manuscripts are that Hindus are the indigenous people of the subcontinent and that they form a single national group. Considering the obvious linguistic, racial, social, and religious cleavages within the Hindu community, Savarkar felt he needed to explain what Hindus had in common that could justify calling them a single national group. British opinion, both official and scholarly, tended to view India as a geographical rather than as a national concept.

Savarkar accepted the notion that the Aryan people and culture had originated in northwestern India and had gradually spread out over the subcontinent. He recognized the existence of non-Aryan peoples; however, he proposed that an intermingling of blood and culture took place between the Aryans and non-Aryans of the subcontinent, producing the Hindu nation. He defines a Hindu as a person who feels united by blood ties with all those whose ancestry can be traced to Hindu "antiquity," and who accepts India—from the Indus River in the north, to the Indian Ocean—as his fatherland (*pitrubhu*)[2]. In addition, a person is a Hindu only if he accepts India as a divine or holy land (*punyabhu*).

While Savarkar's work may have provided Hedgewar with an intellectual justification for the concept of a Hindu nation that embraced all the peoples of the subcontinent, it did not give him a method for uniting the *Hindu* community. Hedgewar had experimented with revolution, *satyagraha*, and constitutional reform, but each method, he felt, had failed to achieve independence or national rejuvenation. From his youth, he had searched for a reason to explain India's inability to ward off foreign domination. He was disturbed that a small group of colonial administrators could rule a vast country like India with such ease. He believed that independence and national revitalization could be achieved only when the root cause of India's weakness was discovered. Some time in 1924-25, he satisfied himself that he had discovered the cause: The fundamental problem was psychological and what was required was an inner transformation to rekindle a sense of national consciousness and social cohesion. Once having created a cadre of persons committed to national reconstruction, he believed there would be little difficulty in sustaining a movement of revitalization, which, of course, would include independence as one of its objectives.[3] The first task on the road to independence, then, was to formulate a discipline and an organization to train the cadre.

The idea of establishing the Sangh became gradually crystallized in Hedgewar's mind. It now only remained to translate that concept into actuality. He chose the sacred day of *Vijaya Dashami,* 1925, for the historic occasion.

That day was, indeed, very auspicious for launching the work, since Vijaya Dashami traditionally symbolized the conquest of good over evil. Founding the Sangh was the first step in the future victorious march of the Hindu Nation—from the individual to collective life, from subservience to invincibility. About 15-20 young men gathered in Hedgewar's house on that day. Chief among them were Bhauji Kawre, Anna Sohani, Vishwanathrao Kelkar, Balaji Huddar and

Bapurao Bhedi. Hedgewar announced to the gathering : "We are inaugurating the Sangh today." He elicited the views of all of them in regard to the activities of the Sangh. "All of us must train ourselves physically, intellectually and in every way so as to be capable of achieving our cherished goal," he declared.[4]

Founding of any institution is usually preceded by fixing up its name, constitution, office, fund collection, etc. There is also a lot of advance publicity. But strangely, there were no such preparations when the Sangh was started. The sole capital of the Sangh at the time of founding was Hedgewar's own life of utter dedication and sterling character, his unsurpassed organizing ability and spirit of self effacing service.

Hedgewar selected the first mission of the young organization with great care. He wanted to demonstrate the value of discipline to both the volunteers and to the general public, and chose a popular religious occasion—Ram Navami—to do so. The Ram Navami festival (celebrating the birthday of Rama) was celebrated in April each year at Ramtek, a village near Nagpur. According to Hedgewar's biographer, the chaotic conditions around the temple created great hardships for the worshippers. Moreover, many villagers were reportedly cheated by Muslim *fakirs* and *brahmin pandits*. He decided to take his volunteers to the 1926 festival to remedy the situation. For the occasion, he chose both the name and the uniform of the organization. The swayamsevaks, in their new uniform, marched to the temple, singing verses from Ramdas. According to RSS sources, they enforced queues for the worshippers visiting the temple housing the main idol, provided drinking water, and drove off the corrupt priests.

Soon after this dramatic introduction to the public, *lathi* instruction and group prayers were incorporated into the RSS discipline. In the same year, a large open area was acquired and military training was introduced by Martandra Jog, a former officer in the army of the Maharaja of Gwalior, who was to become head of the military section of the RSS. This strengthened their sense of discipline, volunteers were required to wear their uniforms to the RSS meetings, and a bugle corps was formed to accompany the volunteers when they marched through the streets of Nagpur. In

1926, the first daily shakha (branch) was held, and the practice of meeting daily was quickly adopted by other RSS groups. Ninety-nine young men were accepted into RSS membership in 1928 by taking a life oath in a forest close to Nagpur. The oath was administered before the *Bhagva Dhwaj*, an ochre coloured standard associated with Shivaji. In RSS ritual, this standard is a symbol of the unity of all Hindus, and it is the "*guru*" to which each *swayamsevak* commits himself when he joins the organization. Some of the older RSS informants recall that the paramilitary orientation of the RSS at that time was popular among the youth of the Hindu middle class in Nagpur because it was proof to them that Hindu young men were the equal in manliness to the British soldier.

Communal riots which erupted again in Nagpur in September 1927 led the RSS to take steps which captured the attention of Hindus far beyond the city. Anna Sohani, a former revolutionary and close associate of Hedgewar, organized RSS members into sixteen squads to protect various Hindu neighborhoods in the city. Perhaps because of the publicity generated by this move, the organizers of the December 1927 Hindu Mahasabha. national conclave at Ahmedabad in Bombay province invited Hedgewar to send RSS members in uniform to the Session.

Hedgewar's revolutionary past and the paramilitary nature of the RSS convinced the Central Provinces Home Department that the RSS could develop into a dangerous revolutionary group, and this suspicion continued throughout the pre-independence period. In fact, the RSS remained scrupulously non-political and it was not until after independence that it began seriously to consider political activities. People who knew the RSS well, such as Dr. Hardikar, the leader of the Hindustan Seva Dal (the youth unit of the Congress) criticized the RSS for its refusal to get politically involved.[5] V. D. Savarkar, the president of the Hindu Mahasabha after 1937, frequently denounced the RSS for its "purely cultural" orientation. In his typically frank manner, Savarkar publicly stated, "The epitaph for the RSS volunteer will be that he was born, he joined the RSS and he died without accomplishing anything."[6]

Hedgewar's stress on the educational role of the RSS—referred to as character building—led some of his senior colleagues, who wanted it to take a more activist stance, to leave the organization. Anna Sohani, one of his closest colleagues, withdrew from the RSS in 1928 when Hedgewar vetoed Sohani's proposal to march uniformed RSS members in front of mosques on Friday as an unnecessarily provocative act. Hedgewar in 1931 condemned the RSS General Secretary, G.M. Huddar, for participating in an armed robbery, even though the money was intended to fund anti-British activities. Huddar drifted away from the RSS after his release from prison. While Hedgewar permitted RSS members to take part in political activities in their individual capacity, he was careful to keep the RSS aloof from them. For example, during his own participation in the 1931 Congress Civil Disobedience Movement, Hedgewar handed over his position to Dr. L.B. Paranjpe, a member of Dr. Munje's political circle, for the duration of his involvement.

Nonetheless, the RSS continued to expand. Because of this growth, Hedgewar called senior RSS figures to Nagpur in November 1929, to evaluate its work and to consider ways to link together the expanding network of shakhas. They decided that the organization should have one supreme guide (*sarsanghchalak*) who would have absolute decision-making power. He would choose all office bearers and personally supervise the activities of the RSS. RSS literature compares the position to the head of a family who acts for the well being of the collective unit.

By a unanimous decision of the senior workers, Hedgewar was acclaimed the first *sarsanghchalak*. While the term *guru* is never employed to describe the position, the *guru* model of authority governs the leadership principle of the RSS. A *guru* in the traditional sense is a spiritual preceptor who knows a path to realization, and his guidance is required for the seekers after spiritual wisdom. A novitiate is expected to obey the *guru* in all matters.

Despite Hedgewar's injunction that loyalty and veneration be focused on the organization rather than on the individual, a charismatic sanctity does surround the office.

Expansion of the RSS: 1931-1939

In the early 1930s, the RSS began to spread beyond its Marathi speaking base in the Central Provinces. Bhai Parmanand, a leader of the Arya Samaj in Punjab, invited Hedgewar to Karachi in mid-1931 to attend the All-India Young Men's Hindu Association session. Taking advantage of the opportunity, Hedgewar started RSS work in the Sind, and soon after launched the RSS in Punjab and, the United Provinces. In the United Provinces, RSS students from the Central Provinces began the work. Among them were Prabhakar Balwant Dani, later General Secretary of the RSS, and Bhurao Deoras, later a senior RSS zonal organizer. The work in north India progressed so well that Hedgewar in 1937 sent ten organizers to expand the organization in Punjab, Delhi, and the United Provinces. RSS informants admit that much of this growth in north India was due to a growing Hindu fear of Muslim paramilitary movements, particularly the Khaskars. Sikh, Hindu, and Muslim para-military groups sprang up all over Punjab in the immediate pre-World War II period. A Home Department Fortnightly Report stated:

"These militant groups continue to multiply. The stock excuse for each successive formation is the alleged certainty of civil war in the Punjab when war is declared in Europe and the necessity for each sect and community to assert itself."[7]

G.D. (*alias* Babarao) Savarkar, a former revolutionary and the elder brother of Vinayak Damodar Savarkar, helped the RSS expand into western Maharashtra. He merged his own Tarun Hindu Sabha (Hindu Youth) as well as the Mukteshwar Dal (Liberation Organization), associated with Pachale Goankar Maharaj (a Hindu saint), into the RSS. He accompanied Hedgewar on trips to western Maharashtra, introducing him to Hindu nationalists. Some of these contacts (e.g., K.B. Limaye, Vinayak Apte, and Bhaurao Abhyanker) were to become prominent RSS officials in Maharashtra. Pune developed into the centre of RSS activities in western Maharashtra.

Between 1931 and 1933, the number of shakhas increased from 60 to 125, and the membership increased to 12,000. Work was started in Madras, Bengal, and Gujarat in 1938, and by 1939 there were about 500 shakhas, approximately one half, in Marathi speaking areas, and 60,000 participants.

A women's affiliate, the Rashtra Sevika Samiti, the first RSS affiliate, was started in October 1936 in the Central Provinces by Mrs. Lakshmi Bai Kelkar, mother of a *swayamsevak*. She sought RSS assistance to fulfil one of the revivalist goals—training women in the martial arts. During a visit to Hedgewar, she reportedly told him, "Just as women are an integral part of the household, so they too are a part of the nation. If the ideology of your organization is taught to women, it would also help the Sangh." However, Hedgewar considered it imprudent for the RSS to accept women. He agreed to assist Mrs. Kelkar to establish a separate women's group. The discipline and organization of this group parallel that of the RSS. While there is no formal connection between the two groups, leaders of the Rashtra Sevika Samiti often consult with their RSS counterparts, and they support the other organizations affiliated with the RSS.

Because the RSS kept no membership rolls during the pre-war period, and because it does not officially recognize caste divisions, it is not possible to make any precise description of its membership. However, RSS informants note that it recruited largely in urban areas and from high caste, middle income groups. Its success in recruiting middle level government employees and teachers worried the government, and a Central Provinces Government Gazette notification in 1932 prohibited government employees from taking part in the activities of the RSS. Nagpur District officials also ruled that teachers in government schools could not join the RSS. This ruling was a result of the 31 December 1931 memorandum from the Central Provinces Department of Local Self-Government, advising local government units that its employees could not participate in "communal" and "political organization."[8] Neither of these restrictions seem to have adversely affected the RSS.

Hedgewar maintained close ties with the Hindu Mahasabha leadership, due to his close association with Dr. Munje and V.D. Savarkar. Dr. Munje presided over the 1927 Hindu Mahasabha annual session at Ahmedabad, and he invited the RSS to perform drills at the session, providing Hedgewar with the opportunity to establish contacts with Mahasabha leaders throughout India. Material at the V.D. Savarkar collection in Bombay suggests that the RSS benefited more from this relationship than did the Mahasabha. Prominent members of local Hindu sabhas wouldintroduce RSS organizers to potential recruits and donors, provide organizers housing and the RSS with a meeting area. This assistance led many members of the Mahasabha, including Dr. Munje, to conclude that the RSS would function as the youth wing of the Mahasabha.[9] Events were to prove them wrong. To emphasize the non-political character of the RSS, Hedgewar refused to sanction RSS support of the Mahasabha's 1938-1939 civil disobedience campaign in the princely state of Hyderabad, though individual RSS members took part in it. Savarkar was trying to turn the Mahasabha into a political party, at a time when Hedgewar was seeking to insulate the RSS from politics. The Mahasabha established its own paramilitary youth group, the Ram Sena, in 1939.

Hedgewar adopted two major strategies for the expansion of the RSS in the 1930s. First, he focused his attention on the universities outside Maharashtra and dispatched his trained volunteers to recruit students in the campuses. The RSS started a shakha in the Benaras Hindu University campus with the help of Madan Mohan Malaviya. M.S. Golwalkar, the successor of Hedgewar, was recruited here. Second, Hedgewar requested the prominent members of the Hindu Mahasabha to introduce the RSS *pracharaks* to the notables of their respective areas and arrange their stay. The Hindu Mahasabha channel proved to be very useful in the initial stage when the RSS was looking for an entry outside Central Province and Berar.

Between 1937 and 1940, the RSS expanded feverishly amidst deteriorating Hindu Muslim relations in northern India. The *pracharaks* were active in Punjab, Delhi, United Provinces and Bihar. In 1939, the RSS entered the southern states of Tamil Nadu and Karnataka. The first attempt to form a Madras branch of the RSS was made in April 1939 when C.S. Paramarth of Nagpur, 'the Chief Organizer of the Madras Province', visited Madras and recruited about 40 members.

The expansion was steady and impressive. There were only

60 shakas in 1931. In 1936, the RSS claimed to have 200 branches and 25,000 members. By 1939, there were 500 branches and 40,000 members; in 1940, the RSS had 700 branches and 80,000 members. Hedgewar also made his organization financially sound. The income of the Sangh came from subscriptions and donations from volunteers, supporters and well wishers, usually collected at the annual *Guru Purnima* festival as *Guru Dakshina.* A 1939 report stated that Rs. 20,000 was kept as a fixed deposit with two Nagpur money lenders, Chitnavis and Ghatate. Strict control was maintained over the finances of the Sangh and each branch was normally permitted to retain only a Rs. 200 deposit.

Along with the expansion and a sound financial position, its character also underwent a transformation from being a regional to an all-India organization. Hedgewar was quite pragmatic; he decided to dilute the Maharastrian emphasis of the RSS to make it acceptable to other regional Hindu traditions. Initially, it did not find an easy acceptance in northern India because of the opposition of the Arya Samaj to the idolatrous rituals and practices of the RSS. Therefore, the worship of Hanuman was given up and insistence on Hindu rituals was less emphasized. The prayer was changed to Sanskrit to break the regional barrier.

Thus, Hedgewar's innovative organizational skill, missionary zeal and perseverance helped the RSS to grow considerably in a short time. By 1940, when Hedgewar died, the RSS had a nationwide presence except Assam, Orissa, and Kashmir. 'I see before my eyes today a miniature *Hindu Rashtra,*' Hedgewar reportedly said in his last speech to the trainees of the 1940 Nagpur Officers' Training Camp (OTC).[10]

The cooling of relations between the Mahasabha and the RSS after Hedgewar's death in 1940 was a continuation of a process that had begun three years earlier when V.D. Savarkar was elected president of the Mahasabha. Savarkar attempted to give the organization a more specifically political orientation. Neither Hedgewar nor his successor wanted the RSS to be closely associated with a group whose, political activities would place the RSS in direct opposition to the Congress.

Savarkar's disdain for Golwalkar further soured relations between the two organizations. Both men were apprehensive regarding the. other's role in the Hindu unification movement. Savarkar did not appreciate Golwalkar's saintly style, and Golwalkar had reservations about Savarkar's unwillingness to compromise. Savarkar's followers, particularly those in Maharashtra, considered him the driving force behind the Hindu unification movement. For example, Nathuram Godse, in his final statement to the court which tried him for the murder of Mahatma Gandhi, states, "Millions of Hindu Sangthanists looked up to him (V D. Savarkar) as the chosen hero, as the ablest and most faithful advocate of the Hindu cause. I too was one of them."[11] RSS members were not prepared to accept this. While many RSS members respected Savarkar, they did not consider him the supreme leader of Hindus. The tension between the two organizations shows up in an angry letter which Savarkar's office issued in 1940 with the advice that "When there is such a serious conflict at a particular locality between any of the branches of the Sangh RSS and the Hindu Sabhaites that actual preaching is carried on against the Hindu Mahasabha..., then the Hindu Sabhaites should better leave the Sangh ... and start their own Hindu Sabha volunteer corps".[12]

After a protracted illness, Hedgewar died on 21 June 1940 in the home of Babasaheb Ghatate, the Nagpur *sanghchalak* (chief executive of the province). On 3 July the *sanghchalak* of the Central Provinces, acting as spokesman for five senior state *sanghchalaks* assembled in Nagpur, announced that Hedgewar had designated Golwalkar his successor on the day before he died. The choice stunned many RSS members, who had expected that Hedgewar would choose an older, more experienced person. Referring to the decision, Balasaheb Deoras, the *sarsanghchalak*, recalls that many RSS leaders were not sure if Golwalkar could handle his new responsibilities.

"Possibly some of us may have thought at that time, that Guruji Golwalkar was new to the Sangh, and not experienced enough. So we might have been doubtful about how he would discharge his responsibility. Those who were outside the Sangh but had love for it were also apprehensive about the Sangh after Doctor Saheb."[13]

Golwalkar's family background training, and interests made him an unlikely choice to succeed Hedgewar. Golwalkar unlike Hedgewar, came from a relatively prosperous and close-knit family. His father was a moderately successful servant who ended his career as the headmaster of a school. Golwalkar, who had a youthful interest in sports and music, had a relatively uncomplicated childhood. His father encouraged him to study the sciences, and he dutifully honoured his father's wishes and was an exemplary scholar. While a student, he displayed none of Hedgewar's intense nationalism; indeed, he was remarkably apolitical. During his late adolescence, he developed a deep interest in religion and spiritual meditation. Two years after earning his M.Sc. in Biology at Benares Hindu University, he was selected lecturer in zoology there. Some of Golwalkar's students encouraged him to attend RSS meetings, but there is no indication that Golwalkar took a keen interest in the organization.

Hedgewar first met Golwalkar while visiting Benares in 1931, and was attracted to the ascetic—twenty-five year old teacher. He was soon to have the opportunity to know him much better. Golwalkar's parents requested that their son return to Nagpur to prepare himself for assuming the duties of a householder. This included studying law, which at that time was considered a promising avenue to higher status. Meanwhile, Hedgewar appointed him *karyavah* (secretary) of Nagpur's main shakha in 1934. In the summer of 1935, shortly after completing his law examinations, Golwalkar was asked to manage the RSS Officers' Training Camp, a clear sign of his high standing with Hedgewar.

Yet, Golwalkar was a reluctant leader. Hedgewar feared that Golwalkar's ascetic temperament could lead him to become a *sannyasi* (a religious recluse). His fears were justified, for in October 1936 Golwalkar abandoned his legal practice and his RSS activities and, without informing either his parents or Hedgewar, left for Bengal to study Yoga under Swami Akhandanand, one of the surviving disciple of Swami Ramakrishna and a colleague of Swami Vivekananda. A RSS pamphlet says of that experience, "...it had also brought to him the realization of 'self' which is the *sin quo non* for knowledge of the eternal and ultimate truth."[14] His religious guide died on 7 February 1937, and a distraught Golwalkar returned to Nagpur where Hedgewar persuaded him to fulfil his religious duties through the RSS. Golwalkar, the filial son, must have had a difficult time reaching the decision to work full-time in the RSS as this meant that he had to ignore his parents' wishes. Nevertheless, he threw himself into his new career with the same energy he had devoted to his earlier academic career, and to religion.

Golwalkar, by his own account, was a rather blunt and short tempered young man, qualities considered inappropriate for an RSS worker. It disturbed Hedgewar. In a rare insight into his own character, Golwalkar recalls that "I had read a vast number of books on various subjects in those days, and there developed around me a thick intellectual sheath through which Doctor's words could not penetrate. But a process of absorption began slowly. I then met him and my whole life changed. I forgot my self-importance."[15]

One could question whether Golwalkar ever fundamentally changed. For example, when he was negotiating the removal of the ban of the RSS in 1948, the government frequently expressed its irritation with his blunt manner. T.R.V. Shastri, an attorney who served as a mediator between the government and the RSS at the time, commented on his abrasiveness: "Mr. M.S. Golwalkar is a blunt man, innocent of the etiquette required in a correspondence with Government. The soft word that turneth away wrath is not among his gifts."[16]

Despite his apprehensions regarding Golwalkar, Hedgewar recognized that the young Madhav had the qualities of leadership. He placed him in charge of the annual All-India Officers' Training Camp at Nagpur for three consecutive years (1937-1939). The position was a key one, and a sign of the respect accorded a person by the RSS leadership. The RSS conducts a large number of camps each year, but the most important are the Officers' Training Camps. Hadgewar, from all accounts, was pleased by Golwalkar's ability to handle complex details of a large camp. Hedgewar was also impressed by his public speaking and literary abilities. In 1938, Golwalkar prepared the first systematic statement of RSS ideology, "We or our Nationhood Defined," a text which he later revealed was an abridgment of an essay on nationalism (i.e., *Rashtra*

Mimansa) by Babarao Savarkar. During Hedgewar's increasingly frequent illness, Golwalkar substituted for him on the speaker's platform. The confirmation of his acceptance into the inner circle of the RSS came at the 1939 *Gurudakshina* festival (an occasion for honoring the RSS flag and contributing to the RSS treasury) in Nagpur, where Hedgewar announced that he had selected Golwalkar to be the new General Secretary, the second most important position in the RSS.

REFERENCES

1. Savarkar, V.D., *Hindutava* (Mumbai: Veer Savarkar Prakashan), 1969.
2. *Ibid.*, pp. 7-9.
3. At about this time Jawaharlal Nehru also expressed dismay over the lack of well-organised cadre of young men which the Congress could mobilise for its various activities. Nehru Jawaharlal, *Towards Freedom* (Boston: Beacon Press), 1958, pp. 120-121.
4. Seshadhri, H.V. (ed.), *Dr. Hedgewar: The Epoch Maker, op. cit.*, p. 78.
5. Palkar, Narayan Hari, *Dr. K.B. Hedgewar* (Pune: Hari Vinayak Datya), 1964, p. 204.
6. Curran, J.A. Jr. *Militant Hinduism in Indian Politics: A Study of RSS* (New York: Institute of Pacific Relations), 1951, pp. 12-13.
7. Government of India, *Home Political File* (1), No. 18.
8. Palkar, Narayan Hari, *Dr. K.B. Hedgewar, op. cit.*, p. 270.
9. Babar, D.E.U., *Changing Political Leadership in an Indian Province: The Central Province and Berar, 1919-1939* (Delhi: Oxford University Press), 1979, p. 106.
10. Kanungo, Pralay, *RSS's Tryst with Politics, from Hedgewar to Sudarshan* (New Delhi: Manohar), 2002, p. 50.
11. Andersen, Walter, K. and Sridhar D. Damle, *The Brotherhood and Saffron* (New Delhi: Vistaar Publications), 1987, p. 40.
12. *Ibid.*
13. *Organiser*, 14 July 1973.
14. *Shri Guruji, The Man and His Mission: On the Occasion of His 51st Birthday* (New Delhi: Bharat Prakashan), 1955
15. Palkar, Narayan Hari, *Dr. K.B. Hedgewar, op. cit.*, pp. 360-61.
16. *RSS Pamphlet, Justice on Trial* (Managlore: Sharda Press), 1968, p. 96.

7

IN SEARCH OF MODEL *

Dr. Hedgewar had realised that the basic weakness of Indian had been their disunity and division on communal, regional and caste lines. He had also come to know that the immortal strength of the people lay in their psyche, their consciousness that Mother India was/is not a mere piece of land, dead and inanimate, but a living entity, having its own body, mind and soul, and a message of its own to deliver to the world, which the Maha-Shakti has written on its cradle.

Now the problem with him was how to eliminate all differences —caste, regional and communal—in one stroke and how to bring the 'Krishna' and the 'Arjuna' of *Rashtra* together at a time and at one place and make them work in the day to day life of the people. He had found out the disease and knew also what the desired medicine was. But the question with him was how to administer it. In other words, how to affect the desired synthesis between the 'Krishna' and the 'Arjuna' of *Rashtra* in the life of the people, or what kind of programme there ought to be or what sort of an organisation there should be to secure the desired union between the 'Body' and the 'Mind', i.e. the 'Kshatra-Teja' and the 'Brahma-Teja', of the people and simultaneously unroll the scroll of the past heritage daily and regularly before them to keep the stream of Cultural Consciousness flowing, and tell them what their *dharma* was in the context of the present in relation with the future, and at the same time wipe off in one sweep all marks of differences the society was infected with. This was the problem with Dr. Hedgewar.

* The chapter is based upon G.S. Hingle, *Hindutva Reawakened.*

He had some models before him from the history of India but they were not to his satisfaction. He studied them one by one but found them incomplete, lacking in one respect or the other, or missing the desired qualifications of the model of his vision in one sense or the other.

Buddha as Model

The first model he studied was the Model which Lord Buddha had tried after the great War of *Mahabharata* when the healthy relations between the 'Body' and the 'Mind' or the 'Krishna' and the 'Arjuna' of *Rashtra* had broken down, setting in a state of complete degeneration in all walks of national life.

It is said that the land had been wiped out of almost all the great souls and the mighty heroes during the Great War. What was left was the mediocre stock, resulting in a state of utter chaos and confusion everywhere.

The national pysche was unhinged from *dharmic* moorings and came to be tied down to many unethical and sick points. It had fallen an easy prey not only to superstitions but to the bewitching lure of Sex and Mammon; it was bogged down in fixed rigidities and moulds and sought to impose the same captivity on life's free and fluid energy. The caste system was inextricably tied with the family tree. *Dharma* was changed into the laws of Machiavellian philosophy; truth was eclipsed and religious corruption had become the order of the day.

The Brahmins had ceased to be seers, the Kshatriyas, warriors and the Vaishas, honest businessmen. In short, when the total psyche of the nation was caught up in formalities, rituals and ceremonies, with hedonistic underpinnings, when the whole social structure was stratified into rigid formulae and conditions, admitting no fresh air, no renovation and adjustment, Buddha appeared on the stage to cleanse the Augean stables.

Buddha found that religion had compromised with the lust for life, subjecting its highest ideals to the satisfaction of the latter's unspiritual demands. It deteriorated into soulless forms, mere externals and priestly obscurantism. It was but natural for him to think of restoring the balance by putting the ethical side of life above the physical side. Therefore, he preferred spiritual culture to physical culture. As the nation had fallen into the pit of moral morbidity and was wallowing in the mire of voluptuousness and sensuality, as the *Artha* and *Kama* were taken to heart at the cost of *Dharma* and *Moksha*, it was quite logical for Buddha to tilt the balance in favour of *Dharma* and *Moksha*. But the irony of the situation was that he tilted the balance far too much in favour of austerity to the point of a total neglect of physical and material culture. In fact, he reversed the movement of the prevailing thought current from body culture to spiritual culture.

Kautalya as Model

The 'Krishna' of *Rashtra* expressed himself forcefully in the personality of Kautalya or Chanakya, the philosopher, and the 'Arjuna' of *Rashtra* in the personality of Chandragupta, the warrior. For the first time in the history of ancient India, the duo, like Krishna and Arjuna of the *Mahabharata* on the battlefield of Kurukshetra, determinedly put their shoulders together, to pull the nation out of the mire it was stuck in. Because they were committed to working together, it was but natural that they should succeed honourably in the long run. As the Lord promised, there reigned supreme peace and prosperity, glory and righteousness for centuries together in the land of Bharata. It is a fact of history that India touched the highest pinnacles of cultural glory during the Mauryan rule, especially during the reign of Ashoka the Great 2709-232 B.C.

Politically speaking, Ashoka's declaration of repentance for the bloody massacre he had committed in the Kalinga War and the subsequent change of his royal policy of conquest by military force to the *dharma* of Buddha, could be considered bad tactics in politics, but culturally speaking, it brought into focus the cultural unity of India—the oneness of the whole of India, the unity which continued to be accentuated and accelerated more and more under other native kings, in spite of the fact that many onslaughts befell the country from the outside world and weakened the national structure.

Buddha had disturbed the balance. Ashoka disturbed it further by overemphasising *Ahimsa*. As time passed the philosophy of *Ahimsa Parmo Dharmah* crept deeper and deeper into the psyche of the people to the point of becoming ridiculous. For example, royal edicts were issued to the effect that drinking water had to be strained before it was given to horses and elephants to drink so that the tiny lives in the water could escape immediate death. Even many corn-throwing-centres were opened in the midst of seas to feed fish in the oceans of the world. But could the big fish stop eating the small fish? Could the greatest attempt on our part to propagate Universal Religion: "Go, ye Bhikkus, to all the ten directions of the world and preach the law of righteousness," eradicate the seeds of animal passion in man? Could the law of Universal Brotherhood ensure India's security and change her sword into rosary? The result was disastrous.

After the fall of the Mauryan Empire, India became an easy prey ground for many of the wild wolves outside the land to come and hunt and kill at will. Many a greedy eye turned to India with covetous hearts and vile desires. Barbaric hordes of the Sakas, the Sungas, the Kanvas, the Parthians, the Kushanas, etc., rushed in to plunder India mercilessly. It was the necessity of the times to re-evaluate forces that were sapping the life blood of the nation.

Ashoka also committed one more mistake and a serious one, and that was imposing Buddhism as the state religion. As this policy ran contrary to the accepted tradition of the land which considered religion as a private affair in which the State would not intervene; it was, therefore, treated very harshly in later days. The people must have revolted against this intrusion of the State into their private affairs and wished honestly to wash this stigma off the land. That is why some historians think that Buddhism was punished so harshly in the land of its birth.

After the death of Ashoka, the Mauryan Empire started disintegrating fast. Under the spell of ideal Buddhism, the 'Arjuna' of *Rashtra* had gone into deep slumber. And seeing how vulnerable the borders were, many a barbaric horde rushed into loot. Those who could not be thrown out began to settle down, creating a very grim situation for the people as a whole. It was at this stage when the 'Krishna' of *Rashtra* became active at once and shortly adopted and Indianised the new comers by putting on them the seal of common nationality. The worthy ones were honourably placed in the Kashtriya Varuna in recognition of their intrinsic worth. Menander or Malinda, Rudradaman, Kanishka and Mahirakula are some of the noteworthy examples.

This miracle of human unification was affected by the magical spirit of this land which Buddha had ploughed with care.

The model that Buddha had tried to resurrect the nation at the time of crisis did not satisfy Hedgewar because it was 'BrahmnTeja' oriented only. It was the appropriate one for the situation it was evolved, but it could not be used under the changed circumstances. When Buddha evolved it, out of the four pillars of life, viz, *Kama, Artha, Dharma* and *Moksha*, the first two pillars, viz, the *Kama* and the *Artha*, were badly affected with the cancerous disease of over love and over indulgence in the objects of sensuousness and animal gratification. It was quite natural, therefore, for the Lord to cut apart the rotten parts and graft in their place the healthy ones, viz, the *Dharma* and the *Moksha*. That was the only solution left with the Lord to secure the life of the bed ridden body of the Samaj at that time. And that he did.

But this model Hedgewar could not put to use because it was one sided. What he needed was a model which could bring the 'Brahma Teja' and the 'Kshatra Teja', i.e., the 'Shakti' and the 'Bhakti' of the people, into play simultaneously and at one place and regularly. Hence his search for the model continued.

The model that Kautalya prepared was complete in one sense but not in the sense which Hedgewar wanted. After the great War of *Mahabharta*, the 'Krishna' and the 'Arjuna' of *Rashtra*, i.e. the 'Bhakti' and the 'Shakti' of the people had joined hands for the first time in the art of nation building in the personality of Kautalya the philosopher, and Chandragupta the warrior. Because (in this model) the 'Brahma Teja' (Kautalya) was backed by the 'Kshatra Teja' (Chandragupta), it was supposed to work; and it did work wonderfully well. Peace, prosperity, victory and righteousness reigned supreme for centuries.

But Hedgewar could not use this model also, because first, it was personality based; it worked because of Kautilya and Chandragupta Maurya. After the fall of the Mauryan Empire, the model stopped working. There ensued a state of degeneration which was hard to stop. What Hedgewar wanted was a long time solution which did not depend on personalities but evolved naturally as the effect of interaction between the 'Shakti' and the 'Bhakti' forces of the people.

Secondly, the circumstances had changed. The kings had left the stage; their system of kingship could not be revived. Nor the third century B.C. sensibility of monarchy could be superimposed on the 20th century A.D. sensibility of Janta based system of government. Hence, the search for the model continued with Dr. Hedgewar.

Gupta Rule as Model

The Gupta period is said to be the Classical Age and one of the golden epochs of Indian history. This was the age of legislators, dialecticians and philosophers. It witnessed the creative and asthetic enthusiasm of the race pouring itself into things material, into the play of senses and the pride and beauty of life. The arts of painting, architecture, dance, drama, all that can minister to the needs of a great and luxurious civic life, received a strong impetus which brought them to their highest technical perfection. It was in this great age that classical sublimities found their marvellous expression in the poetry and drama of Kalidasa, Bhasa, and Visakhadatta, Bharavi, Bhatti, Matrigupta Bhartiretha, Saumila and Kulaputra. Great books like the *Panchtantra*, the *Shakuntalam*, and the *Raghuvansha* were written. Most of the Puranas were updated; the Dharamsutras were codified; the Smritis were given their present form; the Sankhya and the Mimansa philosophies were systematised; the Shilpashastra (Fine Arts), the Kamasutra (Eugenics, Erotics and allied subjects), the Arthashastra and the Shukraniti (Polity) were written. The ancient Indian ideas on Astronomy, Physics, Chemistry, Botany, Zoology, Medicine and Mineralogy were rendered into the treatises by which they are known today.

Aryabhatta, Varahmihira and Brahmagupta were foremost astronomers and mathematicians of this age. The iron pillar at Mehrauli in New Delhi is an outstanding example of the craftsmanship and metallurgical dexterity of those days, and is hard to be surpassed by the modem foundries.

This was also the classical age of cultural synthesis. The solvent of assimilation had become so strong that it dissolved all the multifarious isms and faiths and practically no distinction was left between this god and that god. The cultural life of the nation had become homogeneous. It was a homogeneous culture, a composite culture in the real sense, the culture in which all the composite parts were fully dissolved like sugar in milk.

Now the question arises if many personalities were involved —from Ashoka the Great (274-232 B.C) to Chandragupta Vikramaditya (375-413 A.D) to Harshvardhana (606-647 A.D), and many more big and small Kings, Jagirdhars and Talaqdhars, falling between them, and many artists, designers and planners, who worked, laboured, designed, planned, thought and financed this project—how could they all afford to dissolve their respective egos in the supreme consciousness of the Ajanta Art? How did they remain separate and detached from the sense of involvement from generation to generation and for centuries together without putting their imprint on the walls, frescoes, paintings or anywhere else for the sake of name and fame and to be remembered by posterity? It is a common urge in man to see his name written in golden letters in the history of civilisations for the great works and deeds he has done; but, here is an army of *nishkam karm yogis* who preferred to remain behind the curtain. What could be the driving force which kept them away, aloof and detached, and disinterested? The nation as a whole must have lived in that particular state of mind, a state of aesthetic detachment, which may be called a state of *Bhakti-Yoga-Bhakti* of the Goddess Saraswati, the Mother of all fine arts—and they must have lived in that state of mind throughout for more than 800 years to complete that stupendous and colossal work of Art, that the Ajanta Art is. It would have been *nishkam karm-yoga or Bhakti-yoga*, which must have attracted the attention of the Devas there up in the heaven and made them feel envious over the state of health of the people living here down on earth.

Sometimes, though very rarely, the 'Krishna' and the 'Arjuna' express themselves jointly, in great or small, in individual personalities under special circumstances and for some special tasks. For example, in the Age of Treta there was Lord Rama in whom dwelt both the 'Shakti' and the 'Bhakti' at the same time with a view to killing Ravana, who represented Evil, and establishing the rule of *Dharma* on earth. In *Dwapara* there was Lord Krishna who was both Brahman (Yogi) and Kshatriya (Warrior) in one, who could teach *Gita* as well as wield Sudarshan-chakra, who could give lessons in *nishkam karm-yoga* as well as guide Arjuna on the battle field with the object of vanquishing the Kauravas who stood for the forces of Darkness, and spreading Light of *Dharma* everywhere. In the late ancient period there were Samudragupta and Chandragupta Vikramaditya who possessed both the 'Brahman Teja' and the 'Kshatra Teja' rolled into them with the purpose of purging the land off all the alien rulers and bringing ancient glory to the land of Bharata. And again in the seventeenth century there was born a unique personality, Guru Gobind Singh, who assumed both the swords: the Temporal (Kashtra Teja) and the Spiritual (Brahman Teja), on his person with the object of saving *dharma* and Samaj from the deadly onslaughts of the Islamic Sword. And he created men of steel out of men of clay who cheated the Muslim falcons of their vaunted victory on the battle-field.

Hedgewar believed that such a combination of 'Shakti' and 'Bhakti' in individual personalities occurred very seldom, once in centuries under the configuration of some auspicious stars and that too for some special purposes. He also believed that such a combination, if at all it was going to occur in the Iron Age (*Kali-Yuga*) or say urgently in the 20th Century, it would occur not in the personality of an individual but in the personality of Nation god, i.e., in the collective personality of the people.

By 'collective personality' he meant the general stuff of the society imbued with the positive qualities of head and heart. As the society at large is the product of a social milieu, so the social milieu in that sense may be taken up as the vehicle of expression of the Lord Himself. And the purpose would be to secure Independence first; Independence in all respects, i.e. politically, from the British yoke, economically, from the foreign dominance, mentally, from the western culture, socially, from all sorts of caste aberrations, racial discriminations, regional rivalries, etc., and then establish Bharat Mata as *Jag-Jannani* (Mother of the world) in the community of nations for the welfare of all man, bird and beast.

Hedgewar's concept regarding the Tenth Avatara was not a traditional one—that the Lord would body forth Himself in the form of an individual for the protection of *Dharma* in *Kali-Yuga*. He believed that He (the Tenth *Avatara*) would appear, if at all He is to appear, not in the body of an individual but in the collective body of the people, not in the personality of one specially marked person but in the personality of Virat Purusha, i.e., in the personality of the nation as a whole. Also He, the Lord, would not appear like a flash and go off the stage after dispelling darkness for a few moments, days or years. His appearance would be slow but ever growing slowly and surely growing and maturing and showing His presence in the social milieu. He is born in by stuffing it with the positive qualities of His Head and Heart, and by throwing up great souls after His nature here, there and everywhere to guide the common man on the right path. Out of His Body would come out many *sant-sipahis* (Saint warriors: Saint = Krishna and Warrior = Arjuna; and Saint warrior = Krishna + Arjuna—both rolled into one), blessed with the constructive qualities of His Body (Kashtra Teja) and His mind (Brahmn Teja), who would act as *agresars* (leaders) on His behalf at all the fronts of human life. As a main Power House of the desired social milieu, He would work and light up light houses at every cross road to illumine the path of nation. He would create great souls for all directions and for all fields in the tradition of Rama and Krishna, Guru Gobind Singh and Shivaji, Dayananda and Vivekananda, and go on creating them in series till the KINGDOM OF KRISHNA is established on earth.

Because they (great souls) would work on His behalf and as component parts of His body, so, would they work as true sant-sipahis, brave and bold and fearless, true and honest to a fault, above worldy wishes of name and fame, above favouritism and nepotism, above the machinations of Power Politics and the exploitations of Sex and Mammon, and, above all, the snares of the

Glittering World. Free from crude passion they would be; covetous heart they would not have. Like true light houses, they would monitor the ship of nation safe and sound to its desired harbour.

"Strength lies in collectivity in Iron Age", so goes an adage: *Sanghe Shakti Kalau Juge*. So Hedgewar was in search of that model which could effect unity amongst the different constituents of the nation and at the same time create or help create that worthy social milieu through which the Lord could express Himself and throw up great souls (*sant-sipahis*) in an unending series in all directions till the Mother's work is done on earth. Hence, the search for the right type of model continued with Dr. Hedgewar.

Shankaracharya as Model

The fourth model before Hedgewar was the model which Shankaracharya had tried during the trying days of the nation. On the eve of Shankara's birth the situation in India, politically and culturally, was extremely pitiable. The 'Arjuna' was badly broken; the 'Krishna' was still more badly mutilated. Politically, the northern part of India alone was ruled by 16 small and independent principalities which were at war with each other. Quite oblivious of the fact that their common enemy was stalking on the border to catch them one by one, they were badly engaged in personal warfare over petty affairs. There was no single centralised authority which could bring those divided camps under one banner against the advancing avalanche of Islamic Imperialism.

It was against this background that Shankaracharya appeared on the scene. He tried to re-build the kingdom of Vedic culture by putting the four pillars of life, viz., *Artha, Kama*, *Dharma* and *Moksha*, which Buddha had disturbed, in their proper order. But being tuned to religiousness by birth, his emphasis fell more on spiritual culture than body culture. His approach in the final round turned, more or less, spiritualistic in nature directed towards pure Vedantic life in content and form. Being a spiritual genius, he reduced the number of the existing cults, which were more than seventy two at that time, to six paths, and wove them together on the canvas of *Ekam Sat Vipra Bhaudhā Vadanti* into one piece of symphony called the *Shan-matas*.

Shankara had done a great service to the nation by rehabilitating the 'Krishna', but the body side, the 'Arjuna', he could not rejuvenate. No doubt, under the influence of his spiritual awakening was born later on a very strong Vijaynagar Empire which could resist and stand against the ongoing march of Islamic Imperialism into the South successfully for more than two hundred and fifty years but no viable alternative at the centre was created which could provide an immediate shield against the ongoing onslaughts of the Arabs on the northern side. He was a spiritual guide who could give cultural unity to the land but not a mastermind, like Chanakya before him or Guru Ramdas after him, who could create political unity as well out of the dismantled strands under one centralized authority. Therefore, there did not come up any outward symbol of political unity under his influence. His Digvijaya was only cultural and not political in any sense because in his efforts to rehabilitate what be considered to be the ancient monoistic cult, the Brahmananda of Upanishad, and to counteract the evil effects of the degenerated forms of Buddhism, he stressed the evanescene of the spiritual side of life more and could not, in spite of the deep spiritual character and fervour of his teachings, reveal to the nation its real creative soul. He was not able to discover the point of balance between the 'Krishna' and the 'Arjuna', i.e. between the 'Mind' and the 'Body' of *Rashtra*, which alone could solve the problems of the land. The Brahminical revival, of which he was the moving spirit, meant more than a reorientation of the religious consciousness of the Hindus; it had not in it the power to rekindle the ancient lamp which alone throw light on India's path, to her goal.

Shankara's model was also one sided. Like that of Buddha's, it was primarily *Moksha* and *Dharma* oriented, lacking the physical culture which was also a must for the all round and harmonious people. Hence, the search for the model continued with Hedgewar.

Bhakti Movement as Model

It was history of Medieval India which brought Hedgewar somewhat close to the model of his vision. He knew that the 'Arjuna' of *Rashtra* had never breathed his last throughout the Muslim period. In one form or the other, in more or less degree, he had always crossed swords with the enemy. Not a single day had passed when he had not lifted his sword against the Islamic Imperialism from the day it had stepped on Indian soil. Against Mohammad-bin-Qasim, he lifted in the person of Raja Dahir of Sindh; against Mahmud Ghaznavi, in the person of Raja Bhim Singh; against Mohammad Ghori, in the person of Prithvi Raj Chauhan; against Ala-ud-din Khilji in the person of Rana Ratan Singh; against Akbar, in the person of Rana Pratap, against Aurangzeb in the person of Guru Gobind Singh and Shivaji Maharaj. The Rajputs had their own romance to add to the heroic personality of the 'Arjuna.' "When surrounded on all sides by relentless foes and left without a ray of hope, those flowers of Hindu chivarly and valour silently witnessed their mothers and sisters, wives and daughters leaping into the flames *of jowhar,* and then donning saffron robes plunged headlong with their flashing swords into the enemy forces, never to return. They preferred honourable martyrdom to the despicable existence of defeat and disgrace. The Rajputs had, by such glorious deeds of valour and self-sacrifice, written a dazzling and awe inspiring page of our history. Such flashes of matchless heroism, such a joyful spirit of sporting with death are rarely to be found in the annals of world history."

In spite of all those acts of courage, valour, bravery and sacrifice, Hedgewar knew that the 'Arjuna' was defeated at the hands of the Muslims, not for want of individual strength but for want of collective strength. He had failed in stopping the Ghazis from cutting the heads of the *Kafir* and from carrying away his womenfolk and selling them into slavery and from burning to ashes his hearths and homes, his schools and colleges, his vihars and libraries, his fields and granaries—not that he was weak in armoury, war mechanism, valour, strength or in dedication to sacrifice, but that he was bodily broken into many parts; he had strong arms but they were many and divided. As he could not strike even once with the collective strength under one banner, so was he defeated again and again. The heroes who had smilingly embraced death at the battle fields were the individual sparks and not the collective flame of the Chitti, and so the warrior 'Arjuna' had to fight losing battles most of the time.

Guru Gobind Singh/Shivaji Maharaj as Model

Out of the many stars in the galaxy of the Medieval India, two stars shine most. They shine most because they succeeded in their objective whereas the others had failed or succeeded partially. And they are Guru Gobind Singh and Shivaji Maharaj.

Guru Gobind Singh succeeded because he had assumed on his person both the swords: the Temporal and the Spiritual. In other words, in him were invested both the 'Arjuna' and the 'Krishna' of *Rashtra*, and so we call him "Saint-Warrior". Saint means the 'Krishna' and the Warrior means the 'Arjuna' both rolled into him. He also succeeded in creating an army of saint-warriors (*sant-sipahis*) out of ordinary men of clay, who actually cheated the Muslim falcons of their vaunted victory at the battle fields.

Shivaji Maharaj succeeded because he had a great mother Jijabai, a great Guru Samarath Ramdas Swami Maharaj to play of the 'Krishna' for him and he himself played the role of 'Arjuna', and when the two the 'Brahman Teja' and the 'Kshatra Teja'—met, was assured of peace, glory and victory. And so Shivaji succeeded in establishing Hindu pad *Padshahi* in the teeth of all the heavy odds which Aurangzeb, the Alamgir put in his way. To overcome them was considered a high hill-task in those days but Shivaji succeeded honourably.

It was this one of his prophetic utterances that had come true and it came to be true during his lifetime because there was the 'Arjuna' of *Rashtra* in the person of Shivaji Maharaj to work it out. The long chain of the small bands of soldiers which Samarth Guru Ramdas had prepared here, there and everywhere on the body of the entire state of Maharashtra, served as the first ground-level soldiers of Shivaji Maharaj. Also they acted not only as break waters to the advancing avalanche of the Islamic Imperialism in the Deccan in the initial stages but also served as a standing army later under the banner of Shivaji Maharaj. Moreover, the spirit of Independence, patriotism and sacrifice for *Dharma* and Motherland, which Swami Ramdas had kindled and which Shivaji Maharaj had set to ignition point, not only burnt to ashes the Alanigir's pride and his dream of bringing to knees the Maratha empire but also put to rout the forces of Islamic Imperialism, and continued to glow for a long time under the banner of the successive Peshwas even after the death of Shivaji till the Delhi *Takht* was virtually won.

Hedgewar could not use the model of Samrath Guru Ramdas, i.e., The model of preparing small bands of soldiers here and there on the floor or temple based *Akharas* and of preparing a Shivaji to put them (bands of soldiers) to use against the forces of Islamic Imperialism because of two reasons. First, it was a war-based Model, timing at fighting against the Muslim Power in India (and now the situation was changed). Secondly, it was personality-based, i.e., based on two/three main personalities, say, of Guru Ramdas Swami, Jijabai and Shivaji Maharaj. And when the main Characters left the stage, it began to lose its glowing heat and luster and slowly got itself cooled down.

Under the changed circumstances, Hedgewar believed that the Muslims in India were the Muslims of India. "By changing religion," he opined, "they do not change their nationality. On the pedestal of *Rashtra dharma* they are bound to stand as one with other communities, though at the level of *Vayakti-dharma* they are free to adopt any religion, i.e. mode of worship. In that sense they are as much sons of the Motherland as are the other communities or the Hindus are".

He believed that if Hindu was strengthened first, it would be easy to strengthen the people as a whole. It was the strength of the main flow that would carry the other streams with the mainstream. The greater the strength, the easier the task.

Therefore, what Hedgewar wanted was a broad based model, a model which was based on love for the country and which, by resurrecting the Hindu people, could win over all the constituents of the nation for the service of the Motherland, a model which could work in the society as a whole and for all the times to come and which by injecting the spirit of the 'Krishna' and of the 'Arjuna' simultaneously into the collective body of the people, lift the soul of the Mother to its pristine glory.

1857 Revolution Model

The great Revolution of 1857 was fought by Hindus and Muslims on different grounds. They had their own distinct and separate motives in mind in fighting the British out. The Hindus fought for the sake of Swaraj and *Swadharma*, the Muslims for the sake of restoration of Muslim rule in India.

Hedgewar studied the model in depth and found that 'Arjuna' failed because he had tried to walk on the borrowed crutches of the Mughal Badshah, forgetting that the 'Krishna' of *Rashtra* in the personality of Swami Dayanand had started the Revolution on the implicit strength of the Hindu forces; that he had brushed aside in the very beginning Islamic Imperialism and the 'White Man's Burden Theory' as negative forces in the way of growth of Hindu revivalism; and then on the part of the 'Arjuna' to fall back for help on the Delhi Takht instead of falling in line in total with the Maratha forces under the Kesari Flag of Nana Saheb was sure to invite his death as well as to the Kesari forces and, as a result, death to the Revolution as a whole. And this is what had happened.

Hedgewar learnt that any Revolution in future, if it was to be successful, it must learn the art of walking on its own legs. This was also the lesson which Guru Gobind Singh and Shivaji Maharaj had learnt after the failure of the Bhakti Movement. And, therefore, his search for the model continued.

Vande Mataram as Model

The last model that Hedgewar considered was the model that came into play during the Partition of Bengal. Here the 'Krishna' in the form of 'Spiritual Nationalism' and the 'Arjuna' in the form of 'Militant Nationalism' had come to join hands and fight together against the pernicious design of the British Government to break Bengal. And they fought hand in hand, supporting each other, and won the battle for the first time on the Indian soil against the British. This was the miracle of the two streams: 'Shakti' and 'Bhakti', that jointly flowed from the Gangotri of Vande Mataram and carried along its torrential current the dream of the British government to break Bengal to the bottom of the Bay of Naught.

In 1905 the 'Shakti' of India was invoked by the united voice of the people with the new found *mantra* of Vande Mataram. It was not the physical India; it was her power, her spirit, which was then worshipped by the awakened people. The 'Arjuna' had played the role in the personality of many dare-devil adventurists, i.e., militant nationalists, and the 'Krishna' had spoken in the voice of Lokmanya Tilak, Aurobindo and Tagore and exhorted the people to rise and fight till the Mother's work was done; and innumerable martyrs had come forward to sacrifice their life in the *Yajna* in order to appease the *Mahakal*. For them no sacrifice was too exacting, no suffering too poignant to be endured. Even death could not terrify them, and laughing and singing they had gone to the gallows with Vande Mataram on their lips.

The field of the 'Krishna' (of which Aurobindo, Tagore, Lala Lajpat Rai and Tilak were the chief representatives) was called the field of 'Passive Resistance'; and the field of the 'Arjuna' (of which Savarkar, Hardayal, Avadh Behari, Ashfaqulla and Bismal were the chief exponents) was called the field of 'Active Resistance'—Resistance being the common element of both the fields. Like Krishna of the *Mahabharata*, the 'Krishna' of *Rashtra* would not wield any weapon on the battlefield himself but at the same time would not ask the 'Arjuna' of *Rashtra* to deter from killing the foe with his Gandiva. Also like Krishna of the Mahabharata, he would, always guide the 'Arjuna' of *Rashtra* and devise new ways to defeat the purpose of the enemy.

Likewise, the 'Arjuna' of *Rashtra*, like the Arjuna of the *Mahabharata*, would fight to the last without imparting any importance to any accepted norms of social relationship, ethics of morality or for that matter to any sorts of ethos or morals that came in the way of his singular aim, viz., VICTORY—Victory at all costs, even by adopting dubious means (as the Lord has adopted on the battle field of Kurukshetra) but justified on the grounds of *Rashtra dharma*. "If a demon sits on the breast of my mother," wrote Aurobindo, "and is about to drink her blood, shall I sit idle and coldly calculate whether it is justified or not or I have strength or not to fight? My only duty is to rush to the rescue of my mother". In a similar spirit, he wished, the Indian should approach the political question. "Their primary duty is," he told them, "to save the Motherland. It is for them to rush headlong to achieve the goal, without pausing to think of its pros and cons or of its probable success or failure."

Like Krishna and Arjuna of the *Mahabharata*, the 'Krishna' and the 'Arjuna' of *Rashtra* fought together, supporting each other. Their aim was one and the same, i.e., to defeat the Kauravas, namely, the British Government, in their design to break Bengal. And with that end in view they acted and reacted upon each other; even their actions influenced and widened the scope of the other. It may be safely asserted that but for the newly awakened sense of Militant Nationalism (Active Resistance), the Boycott and the Swadeshi movements (Passive Resistance) could never have developed into a mighty weapon of *Brahmastra* nature. And it is equally difficult to deny that it was the Boycott/Swadeshi movement which brought Nationalism of the fiery nature from the high pitched castles of sentimentalism to the market place of accepted norms of commonsense of a common man. They fought together, and it was the result of their joint adventure that "India moved forward with much faster and longer strides towards its goal in one decade, i.e., between 1905 and 1915, than it had moved in one century, i.e., between 1805 and 1905." And like Krishna and Arjuna of the *Mahabharata*, they won the battle because they had stood together throughout. The partition of Bengal came to be annulled under their pressure. And when they separated again like Krishna and Arjuna of the *Mahabharata*, they lost much of their heat and glowing luster.

When Gandhiji came on the stage, he denounced the cult of violence. Under no circumstances would he approve the theory of 'Bomb and Bullet'. He was hundred per cent pledged to the concept of Non-violence. In other words, he denied totally the right to Arjuna of *Rashtra* to wield his Gandiva on the battlefield. The 'Krishna' was already wedded to 'Passive Resistance' i.e., resistance without his Gandiva. How could he fight without his Gandiva, Hedgewar wondered, and lead the people to victory at the same time? Would Gandhiji be able to lead the country safe and sound to victory on the lone strength of Non-violence while killing the Militant spirit of the

Nation? Could *ahimsa* or non-violence be an unconditional creed and placed above the freedom of India? Should the 'Arjuna' of the *Rashtra* forget the lesson of the *Gita* and bid goodbye to his Gandiva under all circumstances, even when the Kauravas (British) were bent upon molesting the chastity of Draupadi (India)? These were the questions which Hedgewar put to himself and found their answers in the form of a big question mark looming large before his eyes.

Hedgewar was in search of that model which could lift the 'Body' and the 'Mind' of *Rashtra*, both simultaneously, to the heights of self illumination and by infusing new strength and new spirit into the collective health of the people, lead the country to *Purna Swaraj* safe and sound, and at the same time secure a right place for the Motherland in the community of nations where she could play her destined role honourably of *Jag-Janani* in the welfare of all: man, bird and beast :

> सर्वेऽपि सुखिनः सन्तु सर्वे सन्तु निरामयाः।
> सर्वे भद्राणि पश्यन्तु, मा कश्चिद् दुखभाग भवेत्।।

> (Let everyone be happy, let everyone be free from all ills; Let everyone feel good (for others), let none experience the pangs of pain and suffering, grief and sorrow.)

That type of Model Hedgewar was in search of. And so his search for the model continued.

BIBLIOGRAPHY

Aiyar, S.P. (ed.), *The Politics of Mass Violence in India* (Bombay: P.C. Manaktala & Sons), 1967.

Ambedkar, B.R., *Pakistan or The Partition of India* (Lahore: Mustafa Waheed), 1976.

Ambedkar, B.R., *What Congress and Gandhi have done to the Untouchables* (Bombay: Thacker & Co.), 1946.

Andersen, Walter K. and Shridhar D. Damle, *Brotherhood in Saffron* (New Delhi: Vistar), 1987.

Azad, Maulana Abul Kalam, *India Wins Freedom* (Madras: Orient Longman), 1989.

Baker, D.E.U., *Changing Political Leadership in an Indian Province: The Central Province and Beràr, 1919-1939* (Delhi: Oxford University Press), 1979.

Bell Charles, *Indian Mutiny* (London: Clarendon Press), 1931.

Bhambhri, C.P., *Bharatiya Janata Party, Periphery to Centre* (New Delhi: Shipra), 2001.

Bhishikar, C.P., *Deendayal Upadhyaya, Ideology and Perception, Concept of the Rashtra* (New Delhi: Suruchi Prakashan), 1991.

Bhishikar, C.P., *Shri Guruji Pioneer of New Era* (Bangalore: Sahitya Sindu Prakashan), 1999.

Bimal Prasad (ed.), *A Revolutionary Quest* (Delhi: Oxford University Press), 1980.

Bose, S.K., *Tej Bahadur Sapru* (New Delhi: Publications Division, Ministry of Information & Broadcasting, Government of India), 1978.

Bose, Subhas Chandra, *The Indian Struggle* (Calcutta: Published for Netaji Publishing Society), 1948.

Chandra, Bipan, *Communalism in India* (New Delhi: Vani Educational Books), 1984.

Chandra, Bipan, *India's Struggle for Independence 1857-1947* (New Delhi: Viking), 1998.

Chandra, Bipan, *Nationalism and Colonialism in Modern India* (New Delhi: Orient Longman), 1979.

Chaudhary, Sukhbir, *Growth of Nationalism in India*, 2 Vols. (New Delhi: Trimurti Publications), 1973.

Chitkara, M.G., *Converts do not Make a Nation* (New Delhi: APH Publishing Company), 1998.

Chirol, Valentine, *The Indian Unrest* (New Delhi: Lights and Life Publications), 1979.

Chirol, Valentine, *India Old and New* (New Delhi: Lights and Life Publications), 1975.

Collins, Larry and Lapierre, Dominique, *Freedom at Midnight* (New Delhi: Vikas Publishing House), 1997.

Curran, J.A. Jr., *Militant Hinduism in Indian Politics: A Study of RSS* (New York: Institute of Public Relations), 1951.

Dalton, Dennis, *Indian Idea of Freedom* (New Delhi: Manohar), 1982.

Das, Durga, *India from Curzon to Nehru* (London: Collins), 1959.

Das, Swaranmala, *Shri Aurobindo: A Political Thinker* (New Delhi: Ajanta), 1993.

Datta, V.N., *Madan Lal Dhingra and the Revolutionary Movement* (New Delhi: Vikas Publishing House), 1978.

Deodhar, V.N., *Deendayal Upadhyaya Ideology and Perception, A Profile* (New Delhi: Suruchi Prakashan), 1991.

Desai, A.R., *Social Background of Indian Nationalism* (Bombay: Popular Prakashan), 1976.

Dharamvira, *Lala Hardayal and Revolutionary Movements of His Times* (New Delhi: Indian Book Company), 1970.

Dharan, Rajeev, *Justice on Trial* (Manglore: Sharda Press), 1968.

Dignon, Don, *The Indian Revolutionary Problem in British Diplomacy 1914-1919* (New Delhi: Allied Publishers), 1983.

Dawarkadas, Kanji, *India's Fight for Freedom: An Eye Witness Story, 1913-37* (Bombay: Popular Prakashan), 1960.

Gandhi, M.K., *An Autobiography or The Story of My Experiments with Truth* (Ahmedabad: Navajivan Publishing House), 1983.

Ghosh, Kalicharan, *The Roll of Honours: Ancedotes of Indian Martyrs* (Calcutta: Vidya Bharati), 1975.

Ghosh, Partha S., *BJP and the Evolution of Hindu Nationalism from Pheriphery to Centre* (New Delhi: Manohar), 1999.

Golwalkar, M.S., *Bunch of Thought* (Bangalore: Jagaran Prakashan), 1980.

Golwalkar, M.S. and Others, *Integral Approach* (New Delhi: Suruchi Prakashan), 1991.

Gordan, A. Leonard, *Bengal: The Nationalist Movement, 1876-1940* (Delhi: Manohar), 1974.

Gooch, G.P., *Nationalism* (London: Longman), 1945.

Guha, A.C., *First Spark of Revolution* (Delhi: Orient Longman), 1972.

Gupta, Dipankar, *The Context of Identity: Sikh Identity* (Delhi: Oxford University Press), 1996.

Hardayal, *Hints for Self Culture* (Delhi: Rajkamal Publications), 1948.

Hasan, Mushirul, *Nationalism and Communal Politics in India 1916-1928* (New Delhi: Manohar), 1981.

Hasan, Mushirul, *Communal Pan-Islamic Trends in Colonial India 1916-1928* (New Delhi: Manohar), 1981.

Hingle, G.S., *Hindutva Re-awakened* (Delhi: Vikas), 1999.

Jhangiani, Motilal A., *Jana Sangh and Swatantra, A Profile of the Rightist Parties in India* (Bombay: Manaktalas), 1967.

Jog, B.N., *Deendayal Upadhyaya, Ideology and Perception, Politics for Nation's Sake* (New Delhi: Suruchi Prakashan), 1991.

Joshi, V.C., *Sri Aurobindo-An Interpretation* (New Delhi: Vikas Publishers), 1973.

Kakar, Sudhir, *The Colour of Violence* (Delhi: Viking), 1995.

Karan Singh, *Prophet of Indian Nationalism* (Bombay: Bharatiya Vidya Bhawan), 1991.

Karunakaran, K.P. (ed.), *Modern Indian Political Thought* (New Delhi; Allied Publishers), 1962.

Keer, Dhananjay, *Veer Savarkar* (Bombay: Popular Prakashan), 1966.

Kelkar, B.K., *Deendayal Upadhyaya, Ideology and Perception, Political Thought* (New Delhi: Suruchi Prakashan), 1991.

Ker, James Compbell, *Political Trouble in India* (Calcutta: Superintendent Government Printing), 1917.

Kothari, Rajani, *Politics in India* (Boston: Little Brown and Co.), 1970.

Kulkarni, *India and Pakistan* (New Delhi: Sterline), 1998.

Kulkarni, Sharad Anant, *Deendayal Upadhyaya, Ideology and Perception, Integral Economic Policy* (New Delhi: Suruchi Prakashan), 1991.

Malkani, K.R., *The RSS Story* (New Delhi: Sterling), 1984.

Masani, R.P., *Dadabhai Naoroji* (New Delhi: Publications Division, Ministry of Information & Broadcasting, Government of India), 1960.

Masselos, Jim, *Indian Nationalism: An History* (New Delhi: Sterling), 1985.

Mathur, L.P., *Indian Revolutionary Movement in USA* (Delhi: Chand), 1970.

Mazumdar, R.C., Roy Chaudhary, H. C. Dutta, Kali Kumar, *An Advanced History of India* (Madras: Macmillan), 1950.

Mazumdar, S.N., *In Search of a Revolutionary Ideology* (New Delhi: Peoples Publishing House), 1979.

McCarthy, Justin, *History of Our Times,* Vol. II (London: Chatts), 1897.

Mead, Henery, *The Sepoy Revolt: Its Causes & Consequences* (Delhi: Gyan), 1986.

Mehta, Ashok and Achut Patwardhan, *The Communal Triangle in India* (Delhi: Oxford), 1942.

Mody, Homi, *Pherozeshah Mehta* (New Delhi: Publications Division, Ministry of Information & Broadcasting, Government of India), 1963.

Nanda, B.R., *Gandhi, Pan-Islamism, Imperialism and Nationalism in India* (Bombay: Oxford University Press), 1989.

Nath, Shaileshwar, *Terrorism in India* (New Delhi: National Publishing House), 1980.

Nehru, Jawaharlal, *An Autobiography* (London: The Bodley Head), 1958.

Nene, V.V., *Deendayal Upadhyaya Ideology and Perception, Integral Humanism* (New Delhi: Suruchi Prakashan), 1991.

Nisbet, Robert, A., *Community and Power* (Oxford: Oxford University Press), 1968.

Noorani, A.G., *Savarkar and Hindutva* (New Delhi: Left Word), 2002.

Parmanand, Bhai, *Hindu Sangathan* (Lahore: The Hindu Yuvak Sabha), 1936.

Parmanand, Bhai, *The Story of My Life* (New Delhi: S. Chand), 1982.

Parekh, Bhikhu, *Colonialism, Tradition and Reform* (Delhi: Sage Publications), 1999.

Pulkar, Narrain Hary, *Dr. K.B. Hedgewar* (Pune: Hari Vinayak Datya), 1964.

Purani, A.B., *The Life of Shri Aurobindo: A Source Book* (Pondicherry: Sri Aurbindo Ashram), 1964.

Radhakrishanan, S. (ed.), *Mahatma Gandhi* (London: George Allen), 1949.

Rai, Lala Lajpat, *The Story of My Life* (New Delhi: D.A.V. Publications), 1987.

Rai, Lala Lajpat, *Young India: An Interpretation and a History of the Nationalist Movement from Within* (New Delhi: Publications Division, Ministry of Information & Broadcasting, Government of India), 1968.

Rajendra Prasad, *An Autobiography* (Bombay: Asia Publishing House), 1957.

Raju, P.T., *Idealistic Thought in India* (Cambridge: Harvard University Press), 1953.

Ralhan, O.P. (ed.), *Hindu Mahasabha*, 2 Vols. (New Delhi: Anmol), 1997.

Rao, Ramesh, N., *Secular "Gods" Blame Hindu "Demons": The Sangh Parivar Through the Mirror of Distortion* (New Delhi: Har-Anand), 2001.

Ray, Sibnarayan (ed.), M. N. Roy, *Philosopher and Revolutionary* (Delhi: Oxford University Press), 1987.

Reddy, P.L. Sanjeev and Mishra, S.N. (eds.), *Sardar Vallabhbhai Patel: The Nation Builder* (New Delhi: Indian Institute of Public Administration), 2004.

Rolland Romain, The *Life of Vivekananda and the Universal Gospel* (Calcutta: Advaita Ashrama), 1953.

Shankhdher, M.M., *Reflections on Indian Politics* (New Delhi: Kumar Brothers), 1973.

Sarojini, Naidu, *Speeches and Writings* (Madras: G.A. Natesan), 1945.

Savarkar, V.D., *Hindu-Pad-Padshahi* (Pune: Kelkar), 1925.

Savarkar, V.D., *Hindu Rashtra Darshan* (Bombay: Veer Savarkar Prakashan), 1984.

Savarkar, V.D., *Hindutva: Who is a Hindu*? (Bombay: Veer Savarkar Prakashan), 1969.

Savarkar, V.D., *Six Glorious Epics in Indian History* (New Delhi: Rajdhani Granthagar), 1970.

Savarkar, V.D., *Swantantraveer Savarkar: A Guide to Indian Revolutionary Movement* (Pune: Maharashtra Prantik Hindu Sabha), 1963.

Savarkar, V.D., *The Indian War of Independence 1847* (New Delhi: Phoenix), 1947.

Savarkar, V.D., *The Story of My Transportation of Life* (Bombay: Sadbhakti Publications), 1950.

Savarkar, S.S. & Joshi G. M., *Historic Statements: V.D. Savarkar,* (Bombay: Popular Prakashan), 1967.

Sehshadri, H.V. (ed.), *Dr. Hedgewar: The Epoch Maker* (Bangalore: Sahitya Sindu Prakashan), 2002.

Sehshadri, H.V. (ed.), *R.S.S. : A Vision in Action* (Bangalore: Sahitya Sindu Prakashan), 2000.

Shankar, Ramahari, *Gandhi's Encounter with Indian Revolutionaries* (New Delhi: Siddharth Publications), 1996.

Sharma, Jai Narain, *Power, Politics & Corruption: A Gandhian Solution* (New Delhi: Deep and Deep), 2004.

Sharma, Malikarjuna, *Role of Revolutionaries in the Freedom Struggle* (Hyderabad: Marxist Study Forum), 1987.

Sharma Rashmi (ed.), *Sonia versus Vajpayee* (New Delhi: Deep and Deep), 2004.

Shraddhanand, Swami, *Inside Congress* (Bombay: Phoenix), 1946.

Singh, Gulab, *Under the Shadow of Gallows: A Story of a Revolutionary* (Delhi: Rup Chand), 1964.

Singhal, Jwala Prasad, *Sphinx Speaks: The Story of Pre-historic Nations* (New Delhi: Sadgya), 1963.

Sitaramayya, Pattabhi, *History of Indian National Congress* (Delhi: S.Chand & Co.), 1969.

Srivastava, Harindra, *Five Stormy Years: Savarkar in London* (New Delhi: Allied Publishers), 1983.

Suda, J.P., *Main Currents of Social and Political Thought in India*, Vol. III (Meerut: K. Nath & Co.), 1973.

Sullivan, O. Noel (ed.), *Terrorism, Ideology and Revolution* (Sussex: The Harvester Press), 1986.

Tara Chand, *History of Freedom Movement in India* (New Delhi: Publications Division, Ministry of Information & Broadcasting, Government of India), 1962.

The Collected Works of Mahatma Gandhi (New Delhi: Publications Division, Ministry of Information & Broadcasting, Government of India), 1968.

The Complete Works of Swami Vivekananda (Calcutta: Advaita Ashrama), 1994.

Thengadi, D.B., *Deendayal Upadhyaya Ideology and Perception, An Inquest* (New Delhi: Suruchi Prakashan), 1991.

Trehan, Jyoti, *Veer Savarkar* (New Delhi: Deep and Deep), 1991.

Upadhyaya, Deendayal, *Political Diary* (New Delhi: Suruchi Prakashan), 1992.

Varma, V.P., *Modern Indian Political Thought* (Agra: Luxmi Narain Agarwal), 1961.

Vajpeyi, J.N., *The Extremist Movement in India* (Allahabad: Chug Publications), 1974.

Wardlow, Grant, *Political Terrorism Theory, Tactics and Counter Measures* (Cambridge: Cambridge University Press), 1982.

Walport, Stanley, *Jinnah of Pakistan* (New Delhi: Oxford University Press), 1965.

Walport, Stanley, *Tilak & Gokhale* (Bombay: University of California Press), 1962.

Wasti, Syed Razi, *Lord Minto and the Indian National Movement 1905-1910* (London: Oxford University Press), 1965.

Wedderburn, Williams, *Life History of A. O. Hume* (New Delhi: Pegasus), 1974.

Wilkinson, Paul, *Political Terrorism* (New York: Macmillan), 1976.

Zaidi, A. M. (ed.), *The Encycolpaedia of Indian National Congress* (New Delhi: S. Chand), 1979.

Zangwill, Israel, *Principles of Nationalities* (London: Heinemann), 1920.

Index